Kinesis

Kinesis

Exploring Humans in Motion

Dónal Mac Erlaine

Jefferson, North Carolina

ISBN (print) 978-1-4766-9314-9
ISBN (ebook) 978-1-4766-5220-7

Library of Congress and British Library
cataloguing data are available

Library of Congress Control Number 2024002102

On the cover: *top*: "Fencer" by photographer Georges Demeny (assistant to Étienne-Jules Marey), 1906, gelatin silver print (Purchase, Alfred Stieglitz Society Gifts, 2010, The Met); cave art from the Eastern Cape province of South Africa on a farm called "The Valley of Art" (© Photographer/iStock); dancers (© Zeren Yasa/Shutterstock).

Printed in the United States of America

Toplight is an imprint of McFarland & Company, Inc., Publishers

Box 611, Jefferson, North Carolina 28640
www.toplightbooks.com

To my parents

For all its material advantages, the sedentary life has left us edgy, unfulfilled. Even after four hundred generations in villages and cities, we haven't forgotten. The open road still softly calls, like a nearly forgotten song of childhood.

—Carl Sagan

Table of Contents

Preface

We are born to move. Our bodies and minds are made for it. This is a book which explores our relationship with movement, how it manifests in us, how we embody it, and in particular how it has shaped our history as a species. A lot of what is contained in this book treats the issue of how we developed methods of extending our bodies' ways of moving, which have in turn accelerated many developments through time.

The idea of the book came to me during a long-distance running race, when I had an idea (or perhaps I can call it a feeling) of cohesion with all the other runners, racing through the streets of Barcelona. Seeing the other runners all around me in the center of a modern city, I decided to investigate movement as a human phenomenon. So, it was the correlation of movement, the city, and the human mind, which all together formed the territory of this book.

One of the early realizations I had when writing it was that the more ways we have of understanding our history, the better. History is a contemporary phenomenon, composed by those of the times in which it gets written. We can look at our past again and again for it is an infinite well of self-knowledge. This is far truer than the model of a simple straight line of cause and effect running through a course of time.

One of the main positions that I take is that the shift from nomadism to settled living that rapidly happened during the Agricultural Revolution can be seen as a change in how we moved, rather than what we ate. Viewed in this way, we see the first cities that arose as a fundamental disruption of the human as a moving being. The strong connection to place that is so fundamental to human beings today became a huge force in human existence during that time.

Another realization I had was the way humans move after nomadism is discursive. What I mean by this is that we can interpret the way we move, aside from intention, in perhaps the same way as those who hear the discourse of the French Revolution abstractly buried in a Beethoven Symphony.

We didn't actually stop moving, of course, so our kinetic expression was accounted for in other ways. Cities are the force that led us to greatly extend our movement in technical ways by the invention of various vehicles. Living after the Agricultural Revolution also addressed our understanding of movement in spiritual ways (as will be discussed in the latter chapters of the book), relating to the changes in consciousness that followed from these developments.

After agriculture started, it seems the modern human became socially ordered by a new spirituality. This book asks why and how that came to be. This inverse of nomadic movement, living in cities, resulted in two outcomes: firstly, the extension of motion in using vehicular technology; and secondly, the emergence of a desire to overcome movement. This second outcome relates to the connection we have to certain places, and manifested itself in the development of a new spirituality. Both aren't so obviously intertwined, but do co-exist even today, as I shall demonstrate.

The aim of this work is not simply to expand our knowledge, but more to provoke questions and suggest connections between certain events. I do not claim any new discovery, but believe it to be an original work due to its outlook, scope, and perspective. Nor is it a "grand narrative." Rather, I have simply put movement as our primary characteristic. Through understanding the history of our motion, we may see and understand something more to our nature and from there can question the workings and even the meanings that we derive from it.

It is demonstrated throughout that movement is part of our reality. Indeed, it makes up a great part of it. For about 95 percent or more of the history of our species, we were on the move as nomads. Strangely, most of that time is typically considered to be "prehistory" rather than "history"—as history is traditionally considered to begin with civilization. Semantics aside, it is all part of our past, and all can be informative.

As the past doesn't enjoy a material existence, I have taken the liberty of using historical facts in a non-linear fashion. Although of course the laws of cause and effect hold regardless as universal, I have opted not to follow the typical Western linear model. There are temporal jumps, forward and backward when necessary, in the assembly of my exploration. This is not an effort to cherry-pick certain events or phenomena, but rather to illustrate the way I have connected certain aspects of our being. Broadly, it follows the model of using prehistory, then industrial history, and finally reverting back to ancient history, with a helpful dose of colorful anecdotes along the way.

What I present here is non-speculative nevertheless. That is to say it is rooted in sciences such as biomechanics, archaeology, genetics, and urban planning. The background research was broad and delved into everything from ancient philosophy, map-making, geology, history, geography, policy-making, medicine, art, history of ideas, anthropology, and zoology.

Many friends read chapters or entire drafts of this book. Their feedback and criticism were invaluable. Thanks to Angus Whitty, Jason Hunter, James Howlett, Aubrey Ramage-Lay, Dr. Kathryn Harkup, Peter McMahon, Diarmuid Smyth, Dharmachari Vajrashura, Laurel Aroner, Matthew Bastock, Theo Adamou, Dick Walsh, and Begoña Escrig Bengoechea. Thanks to Lisa Camp and Elizabeth Foxwell at McFarland/Toplight for their confidence in the project. Special thanks to Meropi Athanasiou for her extremely close reading, rigorous philosophical criticism, editing, proofreading, and support.

Special thanks also to Francis Matthews whose beautiful artwork adorns the text.

Chapter 1

The Mystery of the River

You cannot step twice into the same river.[1]
—Heraclitus

You cannot step into the same river even once.[2]
—Cratylus

There's a simple element to the design of many stories. It's travel. Travel reflects our deep inner desire to *move.* It's how we physically express our curiosity, seeking out the unknown. It is the driving force of every traditional adventure and it's even been said that "every story is a travel story,"[3] suggesting that travel comes as part of the story's natural arc of tension and resolution. Motion is part of what makes us human. We are moving animals. We are *locomotive.*

That locomotive nature (meaning moving from place to place) relates to everything we are and can be, including the way we move about on two feet, traverse oceans, or even catapult ourselves to the moon. We could even say—as I will in this book—that it is fundamental to to our history as a species. Motion even relates to the development of our modern civilizations and cultures around the world.

The situation where you might find yourself reading this, sitting on the metro or at home on the couch, isn't wholly representative of the norm. Our natural position is not to be almost always at rest, but to be moving a lot of the time. And if we're always supposed to be moving then perhaps we haven't a default "place" to be. We're always between one thing and another, in a flux, never settled. And if it's the case with humans, then why should any other animal be different? It's not as if we're moving about constantly within a world where everything else is remaining stable. Everything is in motion. This was one of Galileo's great discoveries in the early seventeenth century, namely that being in motion (or "inertia") is our natural default state, rather than being at rest.[4]

Heraclitus' River

This is why I chose to head this chapter with what's probably the best-known quote of the ancient Greek philosopher Heraclitus. Born into a wealthy and powerful family in the sixth century BCE, Heraclitus gave up on the possibility of becoming a local king. Renouncing power in favor of ideas, he became a philosopher. Not much of his work has survived, but this idea of constant change was a key ingredient to his thinking. He's telling us that the river's identity changes constantly. We step into it, but the next time we do, it's a different river. New waters are ever flowing.

Heraclitus' aphorism is often repeated in an effort to remind us that life, the universe, and everything are always in a state of flux. Sometimes this is referred to as "Heraclitean flux." And the more we clearly examine the world around us, the more we see that we cannot argue with Heraclitus. And we can even see this *inside* of ourselves. If we were to have a look at our mentality, our thoughts and emotions, we would quickly realize that nothing is ever fixed in the mind either. Those stir about like the weather. Everything always seems to be in a process of change, and nothing ever really sticks.

This constant shifting is true in every realm. Nobody retains their youth, political leaders get voted out, despots are overthrown or die. History is rewritten and newly interpreted with each generation, so even the past appears to be in flux. Heraclitus puts it so simply, describing the passing nature of all things. Fresh water flowing means that the river is more of a process than a thing.

Perhaps the wisdom of change was all well and good, back in the sixth century BCE, on the Western shores of Turkey (then part of the Greek world). That is until a follower of the great philosopher, by the name of Cratylus turned up and made everything much more complex and strange. Today not much is known of him, other than he was a disciple of Heraclitus.[5] And just like his teacher, his thought has come down to us in these "fragments" (this is what scholars call the smithereens of writing that have survived from ancient times). Unfortunately, we can only see these little bits, as if viewing tiny corners and edges of an obscured object.

Cratylus' River

Cratylus seems to be making a joke at the expense of his teacher. But there are some among us who think he was onto something bigger here. His take on the river's changing nature is certainly mysterious—I've quoted it at the head of this chapter too. Aside from his wit, he seems to be pointing to an actual lack of substance in the river, as if it weren't really there at all. Cratylus is denying the existence of the river altogether, a serious claim! Certainly, that is a much more challenging idea than his teacher's, and one foreign to the contemporary Western mind. Indeed, it is more aligned with ancient Eastern thought, as later chapters will explain.

Is Cratylus seriously denying the existence of the river? If so, does he mean to deny the existence of all things? Somewhere in between the two fragments lie the themes and investigations of this book. It is the pattern throughout most of civilized humanity that we tend to look away from what Heraclitus told us, the reality of change. In the crack between these two quotes there lies a great human mystery. We have it described for us in the first quote, and using that, the denial of the existence of the river in the second. And the medium through which we can view this is that of kinesis. (Note that I will use the terms *motion*, *kinesis*, and *movement* interchangeably throughout this book.)

If Heraclitus is right, and all things are always in flux, then it would follow that there is no unmoving river, even glaciers move very slowly. The movement of its waters is the essential truth of the matter. It is kinesis itself that is reality, rather than any fixity, or thing. So, to examine the changing river, we can examine our own motion, that of our species and how and why we move. At the heart of it lies a paradox. On the one hand there is the urge to move (what I call the *kinetic urge*); and on the other is the opposite desire, to rest (what I call the *static urge*). Like the arc of the travel story, it's composed of tension and resolution; travel and arrival; curiosity and acceptance; restlessness and transcendence. Likewise, we have the acceptance of Heraclitus' river and simultaneously the habit of seeing the world as if everything is fixed.

Rather than making any big claims, what this book attempts to do is to go into the creative tension between the kinetic and static desires of the human being and engage with that tension. It does this mostly through philosophy, history, and anthropology—because I found these to be the three most relevant fields of study. I do not attempt to solve anything. Rather, this is an exploratory book. I wondered if we really looked at movement as the primary characteristic of human existence—like many ancient thinkers did as we shall later see—and ask what our history would look like from this perspective, what kind of answers would we get?

Seeing Motion as Culture

Another aspect of motion is how it can be understood as *human culture*. It is as much an expression of ourselves as our languages and music.

The reason for this is that movement is ubiquitous. It's always there, as part of what's going on in our cultural and physical environments. That means that we always have some sort of relationship with it, consciously or not.

My own line of reasoning follows this idea with respect to the culture of our species, the *Homo sapiens.* In all things we are culturally expressive. To see this, we need only remind ourselves of art, music, buildings, and wonderful blends of different foods. And just like the river, we are always in motion. I also investigate the relationship of human motion to Cratylus' claim, which suggests that the river is not there in the first place. So, this book doesn't look just at motion, but our connection to it, our understanding of it. It is a book about culture more than physics.

In today's modern and heavily institutionalized world, there is a tendency to think that culture is typically equivalent to the arts. However, culture is really far more than this, and often it is the type of culture that cannot be classed as art that is more illuminating of our nature as human beings. For example, our languages and the accents in which they are spoken make up culture. Many things that we do even without thinking, that we simply learned as children from those around us are the same. Another example, the use of chopsticks in one place and a knife and fork in another is a cultural phenomenon. This is *not* to say that art is not culture, or that one form is more important than another. Simply put, culture is *human expression.* A painting is no more or less cultural than the accent of the artist who painted it.

I believe that movement can be seen and examined in the way an anthropologist might examine the accent of the artist. Looking at it like this, we see that *motion is culture,* through the myriad species of motion that we engage in—rail, automobile, running, and so on. We could ask questions such as: Is it a political decision to take the bus to work rather than drive a car? How does the outcome of that decision affect our immediate society? What is our relationship with the space that we occupy? Why are we no longer nomads? Because movement is everywhere in our world, there are an infinity of questions like these. In our kinetic culture and physique, we can find the clues that can help answer some of these. The history of our motion reveals why we became so dominant as a species.

It is possible to look at movement from the point of view of the individual, group, nation, or the whole species. I have chosen to see it as an expression of our species in entirety, because I personally value the universals of humanity over the specifics that differentiate us from each other. We can learn more about our deep nature by examining ourselves in this broader way rather than through the hyper-individualism that is so prominent in Western culture today. Ironically, it seems that the more we know about generally being human, the more we can individually be humane.

Physics

Obviously, motion is a physical phenomenon too. We could say that it is the kernel of all physical phenomena. If we were to zoom out and look at a different scale of things, we would see that although it feels as if we can be stationary at times, in fact we're not. We live on a spinning planet, going around its axis at a ferocious speed. The rotation of the Earth, measured at the equator, is about 506 yards per second. In addition to that, it is also whirling about the Sun, clocking speeds of about 18.6 miles per second. Thanks to gravity and a nice cushy atmosphere we don't feel a thing. We are in a system of constant and unceasing kinesis, and it doesn't matter if that perspective is from our place on the Earth, or even our place in the Cosmos. The more we zoom out, adopting ever larger perspectives, the distances and figures become greater and more staggering. Kinesis becomes faster but strangely appears slower.

Looking beyond, as we can do today, we see that our Solar System belongs to a remote arm of the Milky Way galaxy which itself is in rotation about its own center. Movement doesn't ever end, because at some point we reach the limits of the Observable Universe. From what is known in cosmology and astrophysics, it seems that the Earth has never been still, nor has anything else for that matter. Space itself is moving, as it expands, and even the rate of that expansion is accelerating.[6]

Of course, we measure the rotation of the Earth at the equator from a fixed point in space. But how would that be possible? How is any point fixed? Fixed against what, exactly? It could be fixed in relation to the planet's spin, that is one thing. But there really is no fixity to any point in space that is in any way objective, there being no fixed center of anything.

Scaling the other way, down to the atomic and subatomic levels, we find the same thing. All the central parts of the atoms—the protons and neutrons—are surrounded by these whizzing electrons, displaying a striking similarity to our planetary neighbors circumnavigating the Sun. And at the core, the more that we go in, the more things are made up of smaller things, all moving, jumping, and vibrating. This is the infinitesimal world of quantum physics.

It is an implausible realm where all matter can manifest itself as nothing more than waves of energy, with no hard substance whatsoever. Mind-bogglingly small particles such as the atom had been thought of as the building blocks of matter since the times of the ancient Greeks, not long after our heroes Heraclitus and Cratylus.[7] But in the early twentieth century, it was discovered that the outer orbiting parts of atoms—electrons—inexplicably behaved like waves rather than particles in some experiments. It was as unexpected as a tennis ball traveling through the air, turning into a sound, and then turning back into a tennis ball again. It has puzzled some of the greatest scientific minds over the last hundred years, so much so that it's now a given that if someone claims to understand it, it means they don't.

The deeper one goes into this, the more it seems that Cratylus was right. Perhaps there is only energy, and the particles of matter are not really there. They are just like the river to us. We can all agree what a river is, but once we actually attempt to define it, it becomes immensely puzzling. It unravels and ramifies into many philosophical questions. Is the river a body of water holding to a particular shape, or is it a trench naturally cut into the ground? When does a stream turn into a river? If there is no water, is it still a river? And water without a riverbed isn't really the same thing. And so on. Many things become far more complex when we try to examine what exactly they are, and quantum physics must be a leader for ever-deeper counterintuitive facts about matter.

And so, if Cratylus is right and matter doesn't exist, and there is only energy in the form of waves, then there is *only* kinesis. Movement truly is the only reality, even more so than with Heraclitus' river which at least had movement of things. Cratylus is telling us that everything we think is wrong and that there aren't any things in existence. It's not just that everything is moving and spinning, but that there is no "everything," only energy. Perhaps there is correspondence with a new branch of physics known as string theory, which is still only in its infancy today, 2,800 years after Cratylus lived. He's opened up an endless rabbit hole. If there are only vibrations, then they are pure movement. They are pure kinesis without any substance whatsoever.

Why might all this be important? you may wonder. The reason is because if this is the reality of the world we inhabit and move through, then it is fundamental to nature, and we—as humans—are a part of that nature. There's also an element of the spiritual to all this physics stuff, and it has influenced some cultures in their relationship with movement. However, these are big themes and they need their own chapters later.

Because familiarity runs with our immediate environments, I mostly focus on kinesis at the human scale, but looking through both the cultural and physical lenses. We are located somewhere between the cosmic and quantum parts of the scale of things. Our world seems to be one where things exist as objects, flying in the face of Cratylus, but within the Heraclitean Flux. So, we find ourselves somewhere in between the two rivers, we could say. It is a universe of bicycles, ships, and steam trains. We shall look at nomadism, urbanites in cities, transport systems, and some of the revolutions associated with those. We'll also look at the first settling down of our ancestors, and delve into why that happened, and how spirituality played a part in that era of history.

Resistance

For some reason, we are always forgetting the law of constant change. Typically, we don't seem to view the world in the way that Heraclitus or Cratylus are pointing to, even though it's impossible to argue with their positions. Our approach is as if things *are* fixed and unchanging, in continuous denial of the fact that all things change. We invest in properties to be our solid homes forever, we envisage fixed utopian futures, and the news is a constant source of shock as if breaking the mold of fixity. Ordinarily, our world is one where we live as if separate from the rule of change. Perhaps the spiritual philosopher, Jiddu Krishnamurti, had the same river in mind when he noted that "it is not possible to be one with a swiftly flowing river."[8] Our separateness is precisely why we need to repeat Heraclitus' fragment.

Humans tend to think of themselves from the point of view of sedentism, of being still. We can understand constant motion as an idea, but we normally relate to rivers as fixed things. Calling it a river demonstrates this rather simply. Kinesis, then, is something condemned to the study of physics, kept away from our day-to-day lives. It makes sense, too. If we related to the world in any other way, it might make things impossible: there wouldn't be any fixed time or place, and things would only exist as phenomena, or processes.[9] Nothing would have any *thingness* at all and the danger is that we'd live in a world with no values. The thingness, the illusion that we use to view the world does have its merits.

But if we did look at existence in this way—swirling and tossing about amongst the cosmic order—we would certainly understand ourselves differently. Motion can be the bridge between physics and metaphysics (the branch of philosophy that looks at large-scale abstract concepts like being, time, reality etc.). Perhaps even a bridge over one or other of these ancient Greek rivers. We can begin to do this by realizing that movement can be a sense, along with touch and smell etc. From there, maintaining the fact of kinesis as primary, we can enter into a new realm of understanding. One of the few more modern philosophers who examined motion tells us that "movement is reality itself."[10] From this perspective, thingness evaporates, all material things simply come into being and dissolve, as Marx famously said, "All that is solid, melts into air."[11]

Two Opposing Desires

The paradox of human kinesis I mentioned earlier is the tension in this story of humans in motion. Both of these contradictory natural desires—the kinetic urge and the static urge—appear to be universal to all people. The first is a profound desire to move. We are lured by strange places, by strangeness itself even. It's an itch that needs to be scratched, an inner restlessness. It is sometimes a subtle force within us, but nevertheless a powerful one, and not something that can be so easily ignored. There is this insistent continuity of motion. It never ceases. We humans, as part of that reality, don't cease to move either. On the most intimate level, we always have to scratch the itch, and are called to foreign places, hence the element of travel in so many stories. It manifests itself in many varied ways which we shall examine.

For some reason this desire keeps coming back. Just like a real itch caused by an insect bite, the more scratching that we do, the more we need to do. And it leads us to the second aspect of human kinesis: the static urge, a longing to finally scratch the itch for the last time. We must resolve our restlessness. This second desire lies at a deeper level, which we might loosely label as spiritual. We strive to overcome the Heraclitean flux, and the corresponding consistency of movement. Once the tension is resolved in a story, it ends. The hero returns triumphant from his quest, becomes king, and everyone lives happily ever after. Unfortunately, in our real, lived world this manifests itself as us not reacting all that well to change. Predictability in life is just so much easier and less stressful.

In many Westernized cultures, we attempt a life of fixity, seeking out that predictability. The suburban, middle-class dream beckons many. It is often packaged to be the goal of our education, and indeed of our lives. Many settled people see their way of life as positive progress as compared to our hunter-gatherer ancestors. There is a denial of kinesis here happening on some level, ultimately leading so many to find this type of life unsatisfying. The settled suburban life is directly related to a particular species of (non-)movement which we shall revisit later, as it merits its own chapter.

Humans seem to all harbor this inner longing to be inert, to be still. And to do this is to go beyond motion, in some way or other, whatever that may mean. There is a cosmically spiritual quality to this, in how it's expressed. On one hand we have the reality of change, of movement, and restlessness associated with it; on the other the deep desire to overcome it somehow.

Mapmaking

A lovely manifestation of both desires to move and to transcend kinesis can be found in the culture of mapmaking. Maps have been around a long time, the first known ones being produced in the ancient Persian and Greek worlds.[12] Long before the Earth was fully known to any culture, with its continents, islands, coasts, and rivers, curious attempts were being made to map it all out. The unknown parts were simply left unfinished, sometimes with the warning "here be dragons." Today that boundary is where the Observable Universe comes to an end, and the Unobservable Universe begins. When Copernicus published his groundbreaking heliocentric model of the Solar System in 1543, with the Sun in the center, he did so with a map. Maps transcend space, by cramming it all into a single object, representing a birds-eye viewpoint that we can seldom experience. They are strange objects indeed, denying the reality of space and even time by presenting everything at once. Nobody can be everywhere at once, but a map presents itself as if it were possible. The urge to overcome motion altogether is there in a godlike omnipresence.

Nowadays we can think in terms of maps, but there must have been a time when the mind didn't have this learned behavior of picturing space in this abstract way. Like many others, I conceive of space in a map-like way. It exists in my mind's eye, as if looking down from above. Unfortunately, this is very far-removed from the experienced reality of those places when I'm actually there, moving through them. Maps negate movement in one way by capturing everything at once, and in another way, they enable it. They are both kinesis and stasis, contradicting each other so elegantly. They deny movement, but at the same time feed our curiosity about the places they depict. A map whets the appetite of the intrepid traveler. They spur many to move and explore.

Although maps won't be a central feature in this exploration, they do illustrate a human tendency to sometimes attempt to go outside a world of motion. In the broader sense, I attempt to define and look at the culture of motion in an effort for us to understand ourselves as humans in a new way. I divide kinesis into three types, roughly along the lines of physical, political, and spiritual. Unfortunately, it is not a clean linear type of history. History itself isn't a continuous line through time, but something assembled and written after the events themselves, often by the victors. Reality is a little messy at best, so the past that I describe doesn't always follow in sequence as we go from chapter to chapter.

The story of human kinesis deals with physical human motion and the relationship of this to our collective human history. We then discuss the extensions of motion, investigating the machines that are extensions of ourselves: bicycles, cars, trains, and so on. We shall see the consequences that have arisen from their use, and the messages that can be read from these cultures of motion. We will examine the change from nomadic to settled societies, and how we have been "civilizing" space and politicizing it. We shall investigate the *discourse* of movement.

In the final chapters, I deal with the spiritual takeaways of human kinesis—coming full circle, in a way—back to Heraclitus and his contemporaries around the world. We may even solve the puzzle of what Cratylus was getting at! To overcome kinesis means to overcome Heraclitus' reality by deeply absorbing that of Cratylus, no easy feat! It is to fully realize that *being is being-in-movement*, passing from desire to satisfaction, from moment to moment, from birth to death. At the risk of quoting too much from philosophers, a modern thinker opens his book with the question as to what his body does. The answer is "It *moves*. It *feels*."[13]

We humans are embodied, and those bodies that we find ourselves in are meant to move, as the next chapters will argue. The rhythms of the Cosmos maintain an ebb and flow in which we float, tread, and sometimes even swim. So, this form of expression through motion is abstract, spiritual, and ultimately mysterious.

Chapter 2

Dreaming in Footprints

The only given is that upright walking is the first hallmark of what became humanity.[1]

—Rebecca Solnit

Human beings are born movers. Our prehistoric counterparts inhabited a world where the sedentism of the city didn't exist nor even the idea of it. They were always on the roam, perhaps with some sense of a temporary lodging to return to, but probably nothing that matches our understanding of place and rootedness. One of the features which really defines the moving human as a species is their basic and most fundamental medium of transport, walking.

The Miracle of Walking

There exist some startling facts of our physique. We are extremely different creatures from even our nearest relatives, and most particularly when it comes to walking. There are other kinetic aspects to being human, of course, and those shall be investigated later, but our strange, two-legged means of getting about really sets us apart. It seems silly even to say it, but that's only because it is something that we take for granted. It's so much part of what we are that it's not often given much thought.

Of the four great apes (humans, gorillas, chimpanzees, orangutans), all have the ability to move on two legs. For each, this is valid up to a point, and that point varies greatly. Gorillas and chimpanzees waddle a little clumsily, and can maintain this gait for a dozen yards or so. Then they must resort to a rather brutish method of swinging the upper torso from knuckle-anchored arms. Orangutans, who are slightly more distant relatives, are nimble on tree branches and manage to keep their upper limbs free in order to gather fruit. However, they never come down off the trees, so they are really climbers rather than walkers.

Apes aside, most other two-legged creatures hop, rather than walk.

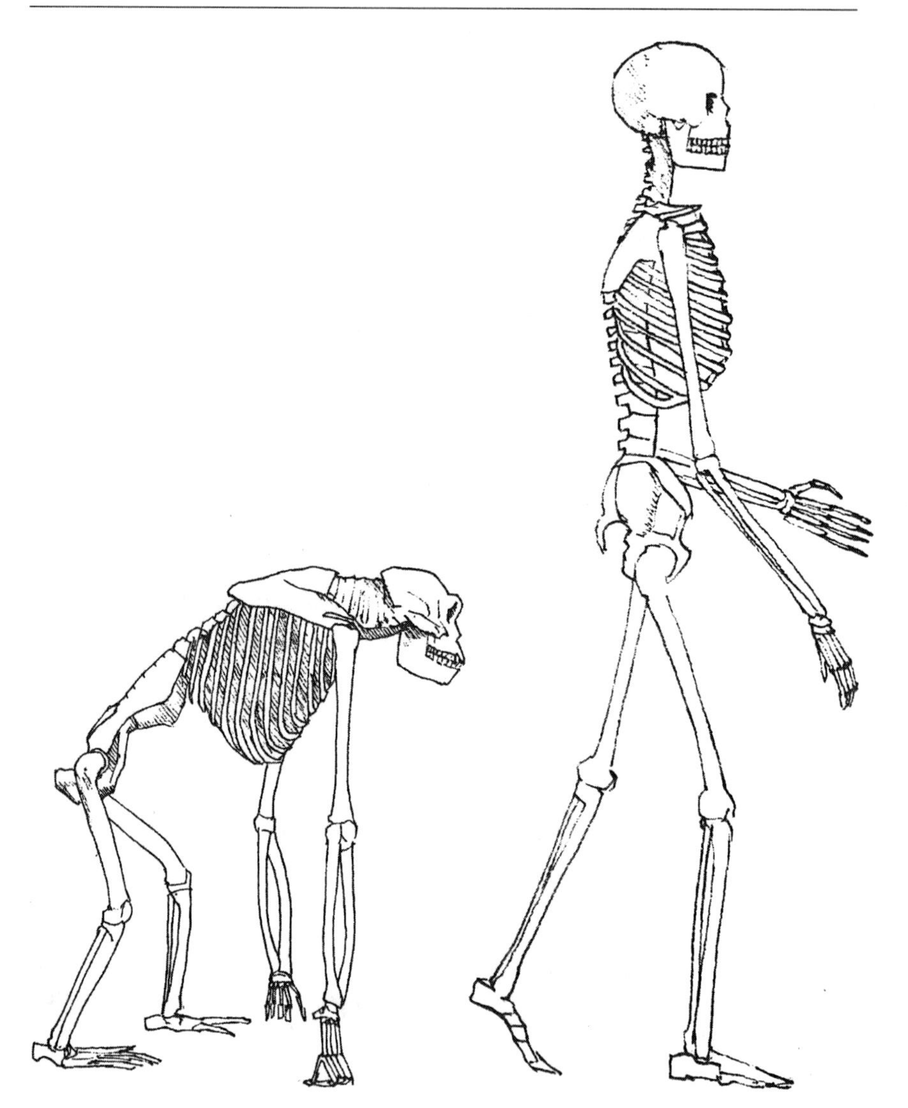

Fig. 1. A comparison of human and chimpanzee. Note the long arms, short legs, and inclined spine on the chimp.

Many bird species fit into this category. Another hopping animal, the kangaroo, is almost entirely made of leg, and hopping appears to be its *raison d'être*. But *Homo sapiens* can glide elegantly on two legs, with apparently no effort made by the arms. Within the whole animal kingdom, bipedalism is already a rare phenomenon, and so the smooth gait of a human being is unique. We might not think of it as such, simply because it's part of our everyday existence.

Much of the reason for this different gait that *Homo sapiens* have is to do with the orientation of the spine. All other bipeds hold their spines with an inclination to the ground, but in humans it aligns itself in a vertical position. In addition to this, humans also have a relatively narrow pelvis and long limbs relative to our weight.[2]

Our legs too, are pretty unique. The other apes do not have the ability to straighten theirs. This means they wouldn't be able to walk for any substantial length of time even if they had the balance. But our long legs go totally straight during the cycle of our gait, and humans use 75 percent less energy than chimps do when walking.[3] This explains how we can walk all day long, and even in today's era of sedentary lifestyles, a reasonably fit person can cover about twenty-five miles in a single day.

Many of our human characteristics that separated us from the other apes are indirectly born out of walking. The vertical spine allowed a total separation of legs from the movement of breathing. In other words, our gait doesn't affect our breathing pattern, like it does in four-legged animals where the lungs sit like a bellows inside the space of the four limbs. This means that a human can talk and walk simultaneously. Therefore, the verticality of the spine made the *Homo* genus more likely to develop language.[4] It was the ability to communicate with such a sophisticated means that enabled complex social systems to arise. This in turn gave early humans the ability to hunt large game, with coordinated hunting tactics in groups.

The complete freedom of the upper limbs, notwithstanding their reduced size when compared to chimps and gorillas, allows the human being to pick up and carry items. Hanging and climbing qualities were reduced, expended as futile. No longer living among the trees, our ancestors were out on the plains. Their arms and hands, although becoming weaker, came to be the creators of tools. The bipedal locomotion of the modern human opened up the possibility for the development of technology by freeing up of hands. Bipedalism defines us as a species more so than opposing thumbs, ability for complex communication, or even intelligence.[5] In terms of our collective biological history, it led to all the other unique cultural characteristics that define us too.

Walking itself is really quite a miracle. In the 1960s, there appeared a seminal research paper about this penned by the British medical doctor and paleoanthropologist John Napier. In this, he claims that "man's bipedal mode of walking seems potentially catastrophic."[6] What he's referring to is the precarious balancing act that happens when we walk. Going from a standstill, firstly both the calf muscles at the back of our lower legs need to relax. This causes the whole body to begin to fall forward, and therefore our center of gravity is on the move. In order not to actually fall, one leg must get forward enough to be ahead of the center of gravity at that time. If it's too far ahead it may stop the momentum generated by the initial "fall," so the calculation is quite precise. This all happens in split-second timing.

Now, in order to propel oneself forward at this point, one must push off with the foot of the standing leg. This is done by the ball of the foot pushing down vigorously, powered by the large calf muscle, and leaving the ground via the big toe. This leg now swings forward just as its compatriot did just a moment ago. If we think of a circle formed by that swinging, the leg being the radius—then the ground would cut into the circle. So, the knee must bend and just enough—too much would cause inefficiency—in order not to hit the ground too early. To explain this is in a way unnecessary, but it illustrates the complexity that is involved. But we can see and experience for ourselves that although Napier is correct in saying that it's touch-and-go, the catastrophe of our bipedalism is only ever potential. We gracefully "wing it" every time we walk. Ironically, the very *instability* of how we walk is what enables our efficiency.[7] Although Napier's paper talks only about the legs, the whole body involves itself in the act of walking. The vast system of muscle and sinew keep the weight of the head well balanced, the eyes level, and the brain protected from vibrations from the ground impact as the legs do their precision work.[8]

Evolution

The oldest evidence of a bipedal ape was found near Lake Turkana in Kenya in 1995. It was discovered by a group led by Meave Leakey, a well-known British paleoanthropologist who was working for *National Geographic* at the time. Named as *Australopithecus anamensis,* this archaic ape-man dates back to just over four million years ago—about one million years earlier than the famous "Lucy," known to us since 1974. From the remains found, we can be sure that *Australopithecus anamensis* walked upright on their two legs.[9]

There are still dramatic physical changes between the *Australopithecus* genus and the subsequent first exponents of what we can call early humans. The earlier of the *Homo* genus appeared in many waves which include *Homo habilis* and *Homo ergaster.* These early people would have scavenged for food rather than hunted systematically. From this we can deduce that the majority of their caloric intake would have come from plants. Systematic hunting would not come about until the existence of later human species (*H. heidelbergensis, erectus, neanderthalensis, sapiens*). The first of these hunter-gatherers with limb proportions similar

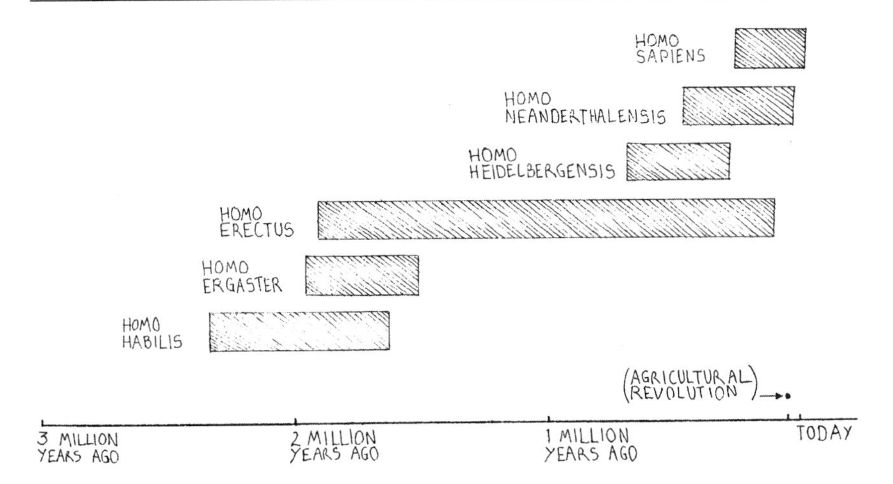

Fig. 2. Timeline of a selection of human species.

to ourselves was to be the most successful human species to date, *Homo erectus*.

Although they were unlike us, we can consider them to be very much a species of human. *Homo erectus* were the first to manage fire and were adept at making stone tools.[10] With the desire to venture forth and explore, they also left Africa and were not without some colonial success as their remains have been found as far away as Indonesia and China. In fact, they were a hugely successful species. If the world were to end tomorrow, it would be *erectus* that would have been the most successful species of human even if they didn't land a man on the moon or develop smartphones. Modern *Homo sapiens* have been around for at least 195,000 years and are already dangerously out of harmony with the environment, while *erectus* lasted ten times as long.[11]

We could consider that all the important things that happened in *sapiens'* history up until the Agricultural Revolution—about 10,000 years ago—happened within the context of walking. That's about 95 percent of our existence. And not only that; it was *because* of our ability to walk that most of these events came to pass, as we shall see.

Dance

One of the behaviors of early mankind which is seen to have contributed to their success is hunting in groups. Perhaps not unlike the murmuration of birds—that graceful appearance of patterns bulging and dissipating within the flock as it moves about—and group movements by other animals, humans are able to connect with each other by tapping into the rhythm of the group.[12] It's a physical way to communicate, not unlike dancing, all the muscular movements of the individual in sync with the others' motion. It has a profound psychological effect too. For this reason, armies have based their fundamental training upon marching drills since the ancient times to unite their recruits, and even during the Second World War it was the cohesive "fellow feeling" developed through drill exercises that kept people fighting, not the ideology or propaganda, according to research at the time.[13]

From an evolutionary perspective, we can consider all group kinesis to have its basis in dance. Because it is a nonverbal type of communication, it can be considered as pure, one which doesn't convey any specific message, like simply looking into someone's eyes. Nothing is said, but there is communication nevertheless. The anthropologist Judith Hanna claims that dance demonstrates that motion has an "inherent value."[14] Furthermore, she goes on to say that "dance is a component of temporal and transcendental behavior."[15] Many have experienced some version of the effects of dance, albeit not a full-blown "trip." There are many examples through the history of religious activity where dance is used for this aim. While the group is rhythmically aligned, the conscious experience of the individual becomes altered somehow.[16] Dance, we could say, is the entry point to trance.

Because this is a feature of all human societies, it can be tentatively assumed that it is somehow intrinsic to being human itself. The importance of group kinesis has led one historian to conclude that large groups and societies probably cannot "maintain themselves without such kinaesthetic undergirding."[17]

Feet Mapping

The typical evolutionary tale of human development is that rather than become more specialized, the modern human became deft at adaptation itself. Having the capacity to adopt learned behavior instead of depending upon genetic mutation means that humans, ironically, are specialized at not being specialized. We are generalists when compared with other species and deal well with change. Although I do not oppose this scientific interpretation, we may say that in terms of kinesis, modern man became specialized in moving on two legs, and their inventiveness later played out in the many extensions of walking and running that we see in modern life.

Standing upright in the long grasses of the open savannah, early mankind developed a new relationship with space. It was a new species, faster than its predecessors, more communicative, and covered ever greater areas. This was to be the species of *Homo* who would dominate the surface of the Earth, doing so on two feet. Understanding that today, with myriad technological inventions surrounding us, the simplicity of their means of this achievement is quite outstanding. Seeking to understand what the world really was, in a way collectively mapping it out for the first time, they ended up in almost every nook and cranny that they could find, over countless generations. This great experiment in exploration is what this 95 percent chunk of human history comprises. Ambulating all over, through space and time, humans learned and recorded the Earth. Returning briefly to the culture of map-making, some cartographers such as Tim Robinson still used their feet in synergy with mind and hands to do the work until recently. It is a fully mindful, even spiritual affair:

> To me, walking is a way of expressing, acting out, a relationship to the physical world; there are of course many others, notably in art. This sort of walking is an intense cognitive and physical involvement ... lingering, revisiting, cross-hatching an area with ones' most alert and best-informed attention. And my maps are the lasting traces of such mobile revises; they are drawn in footprints. Sometimes looking back on the times they represent for me, I feel they have been dreamed in footprints.[18]

Lofty and flowery it may seem, but Robinson's description is close to the reality of some people's experiences. To dream in footprints is the method of recording information among some groups of Australians. What are known as "songlines" belong to a musical song tradition, which incorporates maps into the soundscape itself.[19] The topography and geographical features are literally musicalized into the tonality, lilt, and texture of the music. These nomadic peoples record their means of getting around their vast continent through this medium. Each one is embedded into a story of the "Dreamtime"—a prehistoric time when the land and all animals were created. They are maps drawn with feet and preserved in sound, as if in an ephemeral dance involving both, and quite literally "dreamed in footprints."

Maps aside, Robinson alludes toward walking being the kinetic medium for inner exploration as well as the physical kind, despite the technicalities required by his work. Walking is the simplest and most ancient way to think and feel. It is being human with oneself (when walking alone). We are thinking beings as much as we are walkers, and as one historian of walking puts it, "Since the very beginning, walking and being human have coexisted."[20]

Indeed, some of the greatest philosophical and spiritual minds have been unceasing dreamers in footprints. In ancient times, individuals such as Confucius, the Buddha, Socrates, and Aristotle spent their whole lives wandering about on foot, in a hybrid lifestyle between the settled and nomadic. Aristotle was so well-known for his walking that he was given the nickname *peripatêtikos*, meaning a person who is given to walking (and the root of the word "peripatetic"). These people were nomads in a settled society in a way, and all dedicated themselves to finding meaning in life.

Revolutionary thinkers such as Jean-Jacques Rousseau, Virginia Woolf, and Henry David Thoreau took walking to be a massively important part of their lives. Most of us are familiar with an effect of walking on our mental states. It results in a special type of relaxed clarity, which is the reason why we feel the need to walk during moments of difficulty, as if literally thinking with our feet. Going for a stroll can often broaden our perspective on a problem, and studies have shown that it eases the stress we can feel, living in today's urban world.[21] Rousseau tells us that:

> I am unable to reflect when I am not walking: the moment I stop, I think no more, and as soon as I am again in motion my head resumes its workings.[22]

Along we saunter at our own personal rhythm. As we continue, we become more ourselves, more present and mentally engaged. Sometimes, the seeming pointlessness of walking generates this wider perspective, because it flies in the face of the contemporary attitude that all action must serve an end. Simply walking and being in natural motion rejects the need to always be "doing." From this, a more peaceful mental state arises. Further, there is a spiritual dimension to what walking is and means. This is not only evident in the words of philosophers, but in some religious practices too.

Pilgrimage

From Rome to Bodhgaya to Mecca, the pilgrimage is an ancient and continuing human bipedal spiritual expression. One of the best loved secular (or near-secular) pilgrim routes today is the *Camino de Santiago*. Spiritual investigators and holiday walkers alike come in droves to the small town of Saint-Jean-Pied-de-Port on the French side of the Pyrenees to commence what ends up being roughly a month-long walk for most. Santiago was once the third most important Christian pilgrimage site—after Rome and Jerusalem. Before there was tourism, traveling, and backpacking, people went to "find themselves" through pilgrimage, almost half a million pilgrims per year at one point.[23] Much of the importance of this pilgrimage is related to the journey rather than the destination. It's a nomadic expression, rather than one rooted in a specific place like in Mecca, for example. There, the connection with the place itself is of importance. Mecca draws about one million pilgrims to do the Hajj each year, but few arrive by walking today. These pilgrims—mainly coming from Indonesia, Egypt, and Pakistan—mostly travel by air.[24]

Much of the present-day equivalent to pilgrimage can be found in the public walks that act as fundraising events for charities. These events often bring with them an ambience of healing, community, perhaps suffering, or demonstrating suffering in public. In 1986, the Great Peace March in the USA which was to raise awareness for the campaign for nuclear disarmament, organically became a spiritual event as the walkers gradually became more inclined that way, seemingly because of the walking.[25] There are parallels with the Civil Rights marches of Dr. Martin Luther King and of the Great Salt March led by Mahatma Gandhi.

Historically, pilgrimage was an expensive trip that may have taken years to save for. The links between spirituality, being on two feet, and the cost of pilgrimage was substantially strong enough to give rise to one of the most curious, enigmatic, and beautiful items within all human culture: the unicursal labyrinth. This is essentially a tool for going on a pilgrimage without having to fork out on the usual expenses of travel.[26]

Labyrinths

Contrary to popular opinion, labyrinths and mazes differ in both construction and purpose. A maze is designed to disorientate the participant, causing a puzzle for them to set themselves free once again. For my purposes, by labyrinth I refer to a single path, spiraling in towards the center of the construction, which then continues back outwards, in between the center-bound spiral (hence the term "unicursal"). It is a single path the whole way and leaves no choices. Typically, these fascinating creations, which date back to ancient times, have a high wall just like a maze. But this wall is there for very different reasons in each case. The maze needs a high wall to maintain the disorientation, but the labyrinth uses the high wall to orientate the participant inwards. The only physicality is the awareness of one's biomechanics as they carry on. Labyrinths are tools for self-examination.[27] They provide the possibility of experiencing the conditions of pilgrimage.

For this reason, labyrinths were built into the crypts of many European cathedrals, including one of the most famous, in Chartres.[28] Their origins come from much further back, historically, with Pliny the elder documenting four ancient examples, albeit two had been destroyed by the time he wrote about them. But it was Herodotus before him who documented his visit to the labyrinth in Egypt, which he believed to date to some three thousand years BCE.[29] Not unlike that of Chartres, it was also located beneath another building, a pyramid. Beyond Europe and the Mediterranean, labyrinths are found in many cultures worldwide, from India to North America.[30] We may be familiar with the story of Theseus entering the great maze of Knossos to kill the Minotaur, armed with the ball of string given to him by Ariadne so that he may find his way out. But we are less familiar with the uncountable thousands of pilgrims whose footsteps circled the vaults of European cathedrals through medieval times.

In later chapters we shall discuss how the city can be understood as an expanded version of the labyrinth. The first cities were also built initially for spiritual purposes, places of worship. The walked dreams of those who bridged the single greatest change in human history—from nomadism to urbanism—became internalized, in a way, into the city itself. In later times, purposeless drifting would become a subculture in itself as a way to soak up the contradictions of modernity. The chaos of some cities is perhaps better represented by a maze rather than a labyrinth!

Dance historian Curt Sachs likens the labyrinth to a circle dance, the serpentine twists and turns being ultimately from a common need or spiritual origin. Like in pagan spirituality, the focus is on the immediate human experience, rather than a structured relationship with a higher power. "The essence of the labyrinth is movement," he wrote. "The building has no meaning by itself—it takes on meaning only when people walk through it."[31]

The whole labyrinthine experience is one of easy but focused movement—nothing else takes place other than walking. There is little to nothing to look at, smell, or listen to. The pilgrim walks choiceless in direction and contemplates, encouraged to ditch any intellectual activity, sensing ever less, and allowing a leaning towards self-reflective awareness. The labyrinth is a compressed version of space, and brings with it a similarly compressed experience, a deep experience. It squeezes many important memories, experiences, insights, and realizations into just a few minutes' walking. A long walk in a short space, it's an adventure of contemplative kinesis. Here, walking links spiritual insight to the space within which we exist. One cannot happen without the other. Through their kinesis, the pilgrim becomes one with their own vitality, the bare mysterious fact of their existence.

Perhaps we could consider the labyrinthine experience as better done barefoot, given that there are more sensory nerve endings in the foot than any other body part. Covering our feet, the modern shoe favors speed and efficiency over physical sensation.[32] Without shoes it is easier to get into a mindful state, directly in contact with the ground and our kinetic relation to it. This is certainly an activity which is radical against many of the ideals of modernity, where utility and higher efficiency are valued over mindful awareness. And it is possible to see the humble shoe in this light.

Shoes

Although shoes have pre-dated urban societies dating back at least 10,000 years in areas where people were living nomadically at the time,[33] it was walking in cities which made them the ubiquitous objects they are today. The extent of the change is unclear, but the use of shoes has affected the way we walk over time. Archaeological evidence shows that forefoot striking (landing the foot with the ball rather than the heel) was commonplace as late as the medieval period.[34] Using hard shoes, a walker tends to heel strike, as this is quicker and also more efficient. We have been wearing shoes for far longer than since the Middle Ages, but many of these haven't been the strong, rigid types that we are used to today. Because shoes have more often been made of soft materials like leather, very old examples are extremely rare. But some hypotheses about these changes to the way we walk and how shoes affect us have been backed up by studying the larger bones in the leg. Patterns of arthritis occur more so in the hip before a certain point in time, and in the knee afterwards.[35] The two places are affected differently depending on forefoot strike or heel strike in walking.

Human feet demonstrate remarkable natural engineering, hosting just over a quarter of the bones in the body.[36] We can walk in many ways, and there probably was a wide variety of foot striking, meaning a great variety of shoe types, but once a model was cheaply produced it would have quickly become commonplace.

The invention of stronger shoes meant that we had the rigid sole springing us forward. Some medieval models consisted of a wooden sole atop an iron ring, providing rigidity and some sort of bounce from the metal shape. It's far speedier to walk in a shoe like this than walking barefoot or with soft sock-like shoes. But why would all these changes come to pass, if the human body is so perfect for walking? The answer is simply because we have an in-built preference for inactivity. From as far back as our hunter-gatherer days, we were designed to save energy whenever possible, and the shoe suits this perfectly. In modern history, as life became speedier and cultures grew in the direction of thinking that "time is money," walking slowly was increasingly understood to be a waste of time. Resting trumps, and this marks an important change towards the mainly sedentary life that many of us live today.

Shoes became such successful products that they were adopted in many parts of the world once they became easily affordable. Now they are so widespread, they incorporate information as to how the world is. We could view them as symbols of the modern-day slave trade, being put together in sweatshops in the developing world, meaning that others are literally walking about on cheap or unpaid labor manufactured by those who often have little relation with the products they make. This is ironic given that walking is the mode of transport which is known to "produce" social life, as it typically does not allow for social inequality in the same way that some other transport methods do.[37]

Shoes are associated with spending of high levels of surplus cash in the first world, where the purchase of them has become a pastime for some. In the era of consumerism and male dominance, shoes exemplify our society's values as a whole. They are sometimes a symbol of ostentatious wealth and power—as embodied in the persona of the prominent and controversial public political figure Imelda Marcos who famously owned thousands of pairs.[38] And nowhere is this better demonstrated than in the strangest of human-made objects: the high heel.

High heels are so commonplace in the first world, that it is still an ordinary practice for employers to force female employees to wear them to work, a practice outlawed in very few countries to date. The effects of both making women appear sexualized, and less capable of walking come in tandem with a bizarre twist: the heel as a symbol of empowerment.[39] As one researcher puts it, if they "really did signify power then men would be as willing to wear them as women."[40] And indeed they were worn by men at one point, most famously by Louis XIV of France. This could be said to have been the last time they were really symbolizing power, as the self-described "Sun King" decreed that nobody could wear heels higher than his own. Wearing them demonstrated that one didn't need to walk far, or work in the fields, or run for that matter.

Today, though, they are perhaps the single-most item of modernity which is wrapped up with fetishism (the appropriation of qualities that aren't related to the item itself). The idolism of the bright red high heel, pushed by marketing agencies for years as the symbol of sexual lure and power over men, goes hand in hand with the myth that women are hopelessly addicted to shoes.[41] But in reality, the locus of power is where the subordinate is—in the same way that the sexual attractiveness is really located in the eye of the observer. So, although high heels can certainly be glamorous, sexy, and elegant, the idea that they empower is false. Physiologically speaking, they do the very opposite. High heels cause a shorter stride length, a complete inability to run, and an unsteady gait risking a fall. In addition to this, there is a plethora of other health problems that are related to this deliberate restriction of walking—back pain and foot deformities among them.[42]

The high heel comes not from a walking shoe but the stirrup, morphed out of an invention which revolutionized horse-riding and the warfare of its time—as we shall see in a later chapter.[43] We could say, then, that the high heel did arise out of empowerment (through violence), but in the modern context is subverted into something merely purporting to power via myth.

In conclusion, walking is still the simplest act of human motion despite the cultural complexities that ramify from it. It expresses the countless millennia of physical evolutionary change, the difference between ourselves and those species that came before us, and those who exist alongside us. By walking, we are saying who and what we are—we are demonstrating one of the most unique characteristics of our species. It could be said that to walk is a biologically political act: biological because it is such a part of our physical identity; and political because we communicate things as we walk—be it in a high heel or barefoot in a labyrinth, and particularly so in today's sedentary world. Walking is us stripped down to our bare minimum, moving at a rate of low intensity, just like our ancestors did, ambling and talking. At the same time, walking acts as a vehicle for intense spirituality, propelling the human experience to the limits of understanding. Along each path we walk, hundreds if not thousands have walked before. Their footsteps are countless, leading back to our species' origins. But there was another aspect of bipedalism that set us even further apart from those similar to us: our miraculous ability to run.

Chapter 3

A Unity with Nature

> *I retreat into a universe bounded by my line of sight, the sound of my footsteps.... I narrow the cosmos to this hour, this road, this running. Naught else occupies me. I am content.*[1]
>
> —George Sheehan

Although the majority of us probably don't consider ourselves runners, we may be surprised to find that we don't actually have a choice in terms of this identity. All humans are runners, and that includes those who don't run. How can this be? The answers lie in our physique.

These days, we typically run during childhood far more than in adulthood, and probably because most aren't particularly excellent at it, only the odd child continues with their special talent. Just like children drawing, they do it without much thought, and most love it. Unfortunately, all too often it is only those who are outstanding who are encouraged to continue. The rest of us mere mortals must learn to be content to be part of the hoi polloi. But, when it comes to the subject of running, there is a deeper truth to the matter.

These days running is a "thing," too-often thought of as an activity, rather than being one of the natural ways that we move. In the developed world, humans are increasingly living sedentary lives and compared to our prehistoric counterparts, we are a fitness disaster. Many struggle to maintain an optimum body mass index,[2] and too often find themselves out of breath at the top of a flight of stairs. More than a quarter of the world's adult population today are not active enough, with inactivity at twice the rate in high-income countries.[3] It's not that our ancestors—or indeed present-day marathoners—were incredible human beings; it's just that our present conditioning is one of pampered materiality.

In the previous chapter I mentioned the disposition towards inactivity, a useful tool for early *Homo*, encouraging them to rest and recharge their batteries whenever they were lucky enough to get the chance. We have already seen how it has affected our biomechanics with walking in sturdy shoes. But theirs was a world of far more movement. Too much of our time tends to be at a desk, in a car, or staring at a screen. That natural "laziness" has now become detrimental to our health. Notwithstanding this, the sport of distance running is enjoying a boom.

Nowadays there are commonly held events where people compete over very long distances, such as the marathon (26.2 miles), or ultramarathon (any distance longer than a marathon, some of which are as long as 150 miles). Culturally speaking, it's a special activity and only a minority of humans have done it (about one in every 6,000).[4] As we shall see, much of the current research into prehistory tells us that the *Homo* genus had the equivalent fitness of ultramarathon runners today. In the open grasslands of sub-Saharan Africa where early mankind emerged, they were routinely covering similar distances every few days in order to run down their prey.

There is a feature of our bipedal nature which has been absolutely essential to the survival of our species: our remarkably slow speed. This is far more than simply a by-product of our walking abilities.[5] Human beings are among the slowest runners in the mammalian world. At the time of writing, the fastest known human, the aptly-named Usain Bolt of Jamaica, had recorded a time of 9.58 seconds over the 100 meters (109 yards) at the World Championships in Berlin, 2009. That's a speed of 27.8 miles per hour. But that's the speed of the *fastest* known man, at his athletic peak, in perfect conditions, and sustained not even for ten full seconds. To an ordinary runner like myself it's absolutely astonishing, but being down to only two legs obviously must have its disadvantages. In the world of quadrupeds with whom early *Homo* and *Australopithecus* competed for food it's actually quite poor. Almost all four-legged large mammals can double that speed. Some species of antelope are known to be able to run at 42 miles per hour, and a cheetah holds the record for clocking almost 65 miles per hour, with an acceleration from zero to their top speed in just a few seconds.[6]

Physiology

Why so slow? What sets us apart is a remarkable ability to run long distances. All of these faster animals would die well before the end of a marathon. Today, a little more than one million people run in a marathon each year.[7] The runners' fields include men and women from the ages of about seventeen up to well into their eighties. If you ever watch the first few hundred marathon finishers at a race, something may surprise you—there is no typical body shape. These people are tall, short, round, stocky, skinny—everything. Only sometimes do the leaders of the race have a similar build. The ability to run a marathon isn't just in the physique of the traditional short, lean type. It's in the more generic characteristics of humans.

The amazing piece of engineering that we have at our extremities, the human foot—equally perfect for walking and running—is an ingenious piece of equipment. In *Homo habilis*, who lived from about 1–2 million years ago, there began the development of the plantar arch. This is the spring mechanism which is responsible for the form of the soles of your feet. Modern humans, having far bigger feet, have a more pronounced arch than earlier species.[8] It isn't as important for walking, but it is what powers the push-off from the big toe when running. Just like a stone arch, the more pressure applied from above, the stronger it becomes. It is one of the springing propellant parts of the biomechanics of how we run. In addition to this, it also functions as the first in a series of absorption mechanisms for landing and taking the impact of our moving weight against the ground. Another is in the pronation of the foot, where the ankle bends inwards slightly. Yet another is the bending of the knee and hip when we land. We also have massive lumbar vertebrae towards the lower end of the spine. These joints have very large surface areas in comparison to other primates, which further dissipate the force of impact.

One of the abilities that modern humans have lost is the ability to grip things with their feet. This was forgone probably when the split happened between the *Homo* and *Pan* (the chimp family) genera, about seven million years ago. The "thumb" of the foot has brought itself to be "inline" with the other toes. Having an opposing thumb on the foot is of no advantage whatsoever on flat ground, but having an inline big toe instead can be the lever that the plantar arch of our feet acts upon.

Walking or running barefoot on sand, you may see from the footprints how the placement of the foot is regarding the direction of travel. The big toe, although sitting towards one side of the foot, actually falls dead center. It follows a line, from the print of the heel, through the big toe, following the line of movement. Even the other toes evolved to be a little shorter than they once were in order to cut down on the "metabolic cost" of running.[9] When running on sand, you might notice the footprints themselves coming in line, rather than forming a left-right pattern. They typically follow the line of motion quite strictly.

These days it's common to hear that running is bad for our knees and hips and that the act of running is nothing more than a series of battering impacts on our joints. What this argument forgets is that the feet grip the ground and drag the body *forwards*, not downwards. We bob up and down quite little when compared to how much we're propelled forward. This is further backed up by a scientific study which shows that knee and hip problems are *less* common in runners than in a non-runner control group.[10] The results of another study suggested that it is precisely the long strides—when compared with walking—that saves the runner from the possibility of arthritis.[11] Yet another study took MRI images of middle-aged runners who were setting out to train for their first marathon. They were scanned at various stages of marathon training, and once more after the race. The results showed a *reduction* of damage that was already there due to aging.[12] It seems that the knee joint in particular has the capacity to adapt and maintain its cartilage, as it takes the load over the years, and it's running which induces this mechanism far more so than walking.[13] The ubiquity with which it is believed that running is bad for our knees indicates just how far we've come into the trap of sedentary living.

A further spring mechanism is found in the enormous Achilles tendons and calf muscles. Our legs themselves are unusually long—a trait we borrowed from *Homo erectus*. At the top of our legs, we find the strongest proof of our design being runner-oriented, the *gluteus maximus* muscles. These are the largest muscles in the body, and form the buttocks. We know these to be a runner-specific design because we do not flex them when walking. You can see this for yourself if you simply put your hands on your posterior and walk about. Remaining mostly soft, these muscles don't fire much when we walk, but are fully used when running. They are responsible for that forward drag as the legs swing back.[14]

Some of these facets of human physique do not offer much advantage to a walking human. They instead point to the fact that humans are endurance running machines just as much as they are walking machines. In 2004, a breakthrough scientific paper was published by two American academics, the biologist Dennis Bramble and paleoanthropologist Daniel Lieberman, in which they discuss twenty-six traits of the human body, all of which point towards endurance running. Since this research was published, endurance running has been taken seriously as a defining characteristic of *Homo sapiens*.

There appears to be a continuity in the move away from the *Pan* genus through the australopithecines to modern humans of increased selection for endurance running. The leg of *Homo* is 50 percent longer than *Australopithecus*.[15] The same American scientists found it likely that the Achilles tendon was a characteristic exclusive to *Homo*, and absent in *Australopithecus*. But even with the much longer legs, there is a matching trait of a much slower stride rate in the running human. Even compared to a heavy quadruped, like a lion or tiger, humans have a slower cadence (steps per minute). This makes sense when considering that the legs themselves take up about 30 percent of our bodyweight, because any limb moving in addition to the whole body can be detrimental to the effort put in.

Humans have accordingly developed a more efficient rate of metabolic effort.

As primates we are completely alone in this ability of endurance running.[16] It follows therefore, that it's a phenomenon unique to our biological heritage. It is an ability that separated us from the other great apes, and something we can use to identify ourselves physically and biologically. The more the investigation goes into the connection between endurance running and the human body, the more the two become inextricable. Even our shoulders are low and broad, which provide a stabilizing effect when running.

Skin

A further trait not obviously related to running is perhaps one of the most unique of all: the absence of hair. Of all the 193 known primates, past and present, we are the only ones without fur.[17] Given that it is so unusual, it suggests a specialization in our species, and perhaps for the first time. When the first apes left the trees and landed on flat ground, they were adaptors rather than specialists, but after the epoch of *Australopithecus* and early *Homo*, there come the traits of a running specialist, beginning with *Homo erectus* and our baldness is actually one of the most fundamental.

Humans have three unique characteristics in their skin. The first is the lack of hair already mentioned. A commonly held assumption is that this is because of clothing, but it doesn't make evolutionary sense, as behavior changes before new traits emerge and are selected—a point stressed by Darwin himself.[18] Clothing came far later, long after modern human skin had developed. The second characteristic is a well-developed layer of fat hiding just underneath our skin. This layer of "subcutaneous" fat makes up for the lack of outside fur, forming an ingenious heating system unique to our species. We have our insulation on the *inside*. The replacement of hair by having a layer of fat under the skin allows for the third characteristic: the presence of literally millions of sweat glands all over the body. Again, this is very rare in the mammalian world. The combination of these three traits in the skin allows not only for the ability to run, but for the ability to *keep going*. All creatures begin to overheat during sustained physical effort, but what sets us apart is the ability to offload the excess heat without having to stop, meaning we can continue for quite a long time.

The presence of sweat glands all over the body's surface means that the skin can cool the body. This happens without a hairy interference, making evaporation more efficient. As well as our underlying heating system, there is also an outside cooling system. Sweat cools the skin, offsetting the corresponding overheating. Having a big difference in temperature draws more heat towards that cooler surface. The blood can cool down too, being routed closer to the surface of our bodies, which explains the pink coloring that occurs in the chest or face when exercising. As slow, long runners, we couldn't have better equipment.

Legs and Breath

The position of our legs, in the context of the whole of our bodies has a big part to play in the modern human. Once an antelope, say, has broken out of a light trot into running proper, its front legs pull all the way backwards, while the hind legs overtake them to complete the stride, its body opening and closing as it gallops along the savannah. Within the space of the four legs is the mechanism which constantly replenishes the energy to run: the lungs. With each stride, the lungs are compressed and then expanded. What this means is that only one breath can be taken per stride. The breathing of a four-legged animal is inextricably linked to the gait of its running.

Humans, on the other hand, are able to control—with absolute precision—the rate of breathing and the speed of running as if they were two unrelated actions. We really only need to think about one, as the breathing can work away by itself. Our bipedal design keeps the lungs totally separate from what our legs get up to. So, we have the ability to fine-tune our running speed, meaning that we can move with greater efficiency over distance. This also means that humans can run with a slower cadence, for the speed that they do go at, as already mentioned.

The antelopes and cheetahs I mentioned at the beginning of this chapter can easily outrun us in a flash, but at some point, they need to stop and recover after the sprint. They do this by panting, making up by rapidly refilling with oxygen, and taking a chance to cool down after the intense effort. Many of the species that our early ancestors hunted and fed upon did exactly this—run fast, stop, and breathe very heavily for a few minutes, and then rest. Otherwise, they'd be in serious danger of collapsing from exhaustion. A galloping mammal cannot pant while maintaining more than a trot.[19] *Homo*, hunting in the plains and grasslands of Africa, could keep going however, while still breathing heavier than normal. Humans can breathe through the mouth to meet the increased demand for oxygen, whereas other apes breathe through the nose almost all of the time, meaning there is a higher level of resistance when breathing heavily.[20]

They weren't able to take advantage of the separation of lungs from legs like *Homo* did.

So how did *Homo sapiens* manage to outrun the faster animals? Patience! Their slow and continuous running meant they gradually caught up with their prey as they had to stop and recover from their sprinting efforts. Once the gazelle or kudu would see them coming, they'd take off again, repeating this pattern until they finally collapsed in exhaustion.

Compared to *Homo erectus* and *neanderthalensis* who were also good runners, inventive with tools, and perhaps somewhat communicative, the early *sapiens* had anatomical features which gave advantages over the others. One of these was the narrower pelvis, more efficient for running than the broader shape that *erectus* and the Neanderthals had. Another was the width of the ear canals. This is a crucial part of the anatomy when it comes to balance, particularly balance whilst in motion. *Sapiens* had larger ear canals and therefore a better sense of balance and agility. We could run, jump, and clamber at speed while able to rely on our abilities to land well with the following step far better than our Neanderthal cousins. The instability of our posture which allows us to walk so efficiently—mentioned in Chapter 2—is used fully when we run. When walking, our weight is about 80 percent supported by the active leg, but with running this stretches all the way up to 100 percent.[21]

Because of the ability of almost full rotation of the shoulder in *sapiens*, we also had the kinetic ability to balance the head very well while running.[22] This also gave us the ability to throw a spear, whereas the other species of humans couldn't. Hunting on the plains, this was a gamechanger. *Sapiens* were extending their reach of motion beyond the body, through this basic technology. The Neanderthals could not throw missiles which meant they had to enter into close combat with their prey using their "thrusting" spears.[23] This may explain why they subsequently survived for so long in the woodlands and forests of Europe, occupying that area well before modern humans arrived there.

Behavioral characteristics such as highly sophisticated language skills gave us an edge when it came to competition for food. A complex system of communicating information is crucial to learning how to track animal footprints in the savannah, so the development of sophisticated language was key to our survival. That ability to communicate is what allowed the modern human to hunt in groups. Arranging and coordinating themselves in hunting bands which covered large areas, early humans were able to catch the faster game. For example, agreeing a set of hand signals in advance can allow hunters to work together in absolute silence so as not to scare away their prey. This all depends on the culture of invention that is so unique to humans and which we see manifest in so many ways today, as creative beings. One genetics expert says that "human culture feeds into itself, thus generating its own accelerating tempo,"[24] and our social and cultural nature today stems from this dependency of being part of a group.

Hunting

Today, a group of hunter-gatherers known as the San are using one of the most ancient hunting methods known. They reside across a huge sub-Saharan area encompassing Botswana, Namibia, Angola, Zambia, Zimbabwe, South Africa, and Lesotho. Some of their hunting expeditions can last hours or even longer, bringing them a full day away from their villages at times. Their method of hunting is thought to be close to that of early humans many millennia ago.[25]

Persistence hunting happens normally right in the middle of the day, during the siesta hours. The advantage of being able to tolerate high temperatures without stopping means that people can run down their prey precisely when this makes all the difference. Most large animals in the African grasslands take refuge from the heat during the hottest hours of the day, and so humans had a slight edge during the afternoon. The difference between the abilities of predator and prey is often very slight.

Hunting often comes with a somewhat romanticized image of men hunting for meat while women forage and gather, but there is evidence to suggest that this picture is a misleading simplification. Experts claim that San gender divisions do not adhere to this idea of "man the hunter," and "woman, the gatherer" as they put it,[26] and there are accounts of !Kung men—one of the best-known of the San peoples—doing the hunt within a matriarchal society (the exclamation point resembles a clicking sound).[27] Anthropologist Patricia Draper states that sex tends not to be a strong basis for separation in these societies (like it can be in Western society, from which the San are often studied). She says that most of the experts on a large-scale Harvard project to study these people agree that the "!Kung society may be the least sexist of any we have experienced."[28] This is something that set *sapiens* apart from Neanderthals—who seemed to have had very strong gender divisions with regard to hunting.[29] Egalitarian hunting practices are seen in other hunter-gatherer groups today, such as the Aka who live in Central Africa.[30] Further afield, remains of women hunters have been found in the Americas suggesting that women's role was "nontrivial."[31]

What may egalitarian hunting practices have to do with running? Well, the answer could be found in some simple arithmetic. There is an unsuspected phenomenon to be found when we look at distance running in terms of gender. Women are typically slower compared to men by about 10–11 percent over short distances such as the 100m, 400m, or 800m. For example, the current women's 100m record (at the time of writing) is that set by American sprinter Florence Griffith Joyner in 1988, a time of 10.49 seconds—a difference of about 11 percent to Bolt's 9.58. This difference continues as a constant percentage over short and middle distance, but it *shrinks* once they are into the longer competitions. One recent study examining a large number of ultramarathon finishing times demonstrated that at a certain point, that percentage actually shrinks to less than zero. At the 100-mile race distance (161km), men and women are almost neck and neck, men being only 0.25 percent, and as the distances get longer, women *pass out* their male competitors. At any distance over 194 miles, women are faster.[32] Women have also been shown to be far better at pacing than men—about 18 percent better in fact—and every distance-runner knows the mantra: "it's the *pace* that kills, not the distance."[33]

There is an aphorism stating that every morning a lion wakes up knowing that it must be the fastest, otherwise it shall go hungry. And every morning an antelope wakes up knowing it must also be the fastest, otherwise it won't live through the day. Each creature operates at the boundary of its capability. Although it is at the limits of comfort for a human to run on the African plains at two o'clock in the afternoon, it's *possible*. It's the tiny difference of the slowest and fastest in the herd which makes the kill. If it wasn't like this, the slower species (i.e., the food) wouldn't survive, and neither would the faster species have anything to eat. So, the difference in speed among animals holds itself in a careful balance for both to survive. In kinesis, nature is delicate, even when considering the violent actions of a lion attacking its prey. This difference also exists between *sapiens* and their close relatives. The new human species only needed to be a fraction faster than its competitors to catch the game.

In southern Africa as the San hunt, they operate in silence, with plenty of communication between the various participants through hand signals.[34] This minimizes the disruptive noise that might frighten their prey. The long days of running and tracking in silence that they perform was once described by David Attenborough as achieving a "trance-like state of concentration," which is then followed by a spiritual ceremony of respect and gratitude when the animal is finally killed by its hunters.[35] And any reader who has some experience in running will be somewhat familiar with this state of mind. Although it's more associated with longer distance running, it is a phenomenon that can affect anybody. Most likely, it developed this particular association with marathon and ultra-marathon distances simply because of the longer amount of time spent at it meaning it's likely for those runners to experience it.

There is a clue to the "trance-like state" and the effort expired. It is to be found in our metabolic effort during this task. In their celebrated paper, Bramble and Lieberman plotted the relationship between metabolic effort against speed. This shows the changes between efforts required to maintain certain walking speeds, compared against other animals. They all plot a "U" shape, which is to say that going very slowly is inefficient; a little faster is more efficient, until the curve follows its U-shape back up again as the speed is increased. There is a sweet spot for efficiency around the lowest part of the curve. But for a distance-running human—running between approximately five and ten miles per hour—the curve is almost a flat line! This correlates with the trance state—often remarked upon amongst marathon runners too[36]—presumably because it's an energy saving measure. There is an increase of only 30 percent in energy expenditure with running when compared to walking, but considering the ground covered, it is a vast difference.[37] Remarkably then, running can be *more efficient* than walking.[38]

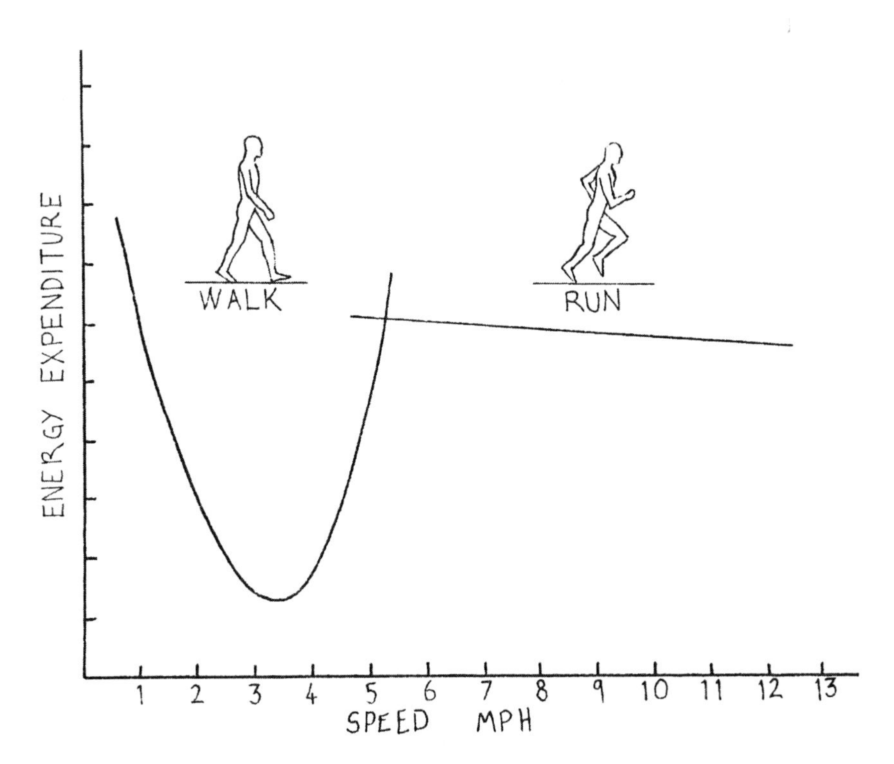

Fig. 3. Energy expenditure (calculated by a metabolic cost per weight, per unit of distance) compared to speed in walking to running, from Bramble and Lieberman.

One of the other major contributions to world culture that have come from the San is their rock art. Some examples include running and dancing. Much of their art depicts hunting, experiencing trance, and running.[39] This is where all three come together, in a culture which is also responsible for the oldest example of modern human culture, some 73,000 years.[40] These are regarded by some experts as "deeply spiritual art."[41]

Fig. 4. Examples of San rock art, depicting trance-like states induced by dancing (above), and persistence-hunting runners (below).

Flow

The world-famous track athlete Roger Bannister experienced this trance state during a run of just a few minutes. In 1954, when he became the first athlete to break the threshold of the four-minute mile—a feat thought impossible for years—he had this to say about the experience: "No longer conscious of my movement, I discovered a new unity with nature. I had found a new source of power and beauty, a source I never dreamt existed."[42]

There is a glimpse of something of the beyond in this quote, something extraordinary in how he mentions a "unity with nature" while he's "no longer conscious" of his motion. It was not until 1990 that the Hungarian-American psychologist Mihaly Csikszentmihalyi defined this state as "flow." He specifically mentions running as an activity which can produce an "optimal experience," by which he means a temporal experience where we are at our best in terms of happiness.[43] One of the major conditions necessary to achieve this state is engaging in an activity which he defines as "autotelic." This is when we do something without a goal in sight, doing it simply for its own sake, and where we are immediately rewarded.[44] Running, in the modern context where we aren't chasing down our dinner, is such a condition. Of all the other activities which are mentioned, there are none which don't involve movement, other than meditation.

Running in a flow state can feel like a sort of oneness, where there is no split between the mind and body. One *is* the other. It isn't something which is easy to describe, but anyone with some distance-running experience may understand it. It's a feeling of naturalness, as if one's true humanity is being kinetically expressed. We abandon ourselves in the goallessness, thoughtlessness, and repetitiveness of running. In the words of biologist and ultra-marathoner Bernd Heinrich, "There is nothing quite so gentle, deep, and irrational as our running—and nothing quite so savage, and so wild."[45]

Running Monks

Far away from southern Africa, the flow state has been institutionalized on Mount Hiei, in Japan. Here, for many centuries now there has been a tradition of running monks. These men are Tendai Buddhists, and have been doing this as a spiritual practice since 805 CE, when the founder of the school, a monk by the name of Saichō, established a temple there. The most common form of run for the *gyōja* ("marathon monk") is a 100-day workout, in which he runs a distance of about forty-five miles through the night, and repeats this each night for ninety-nine nights. If this is not enough, there is the option of a seven-year version of the same thing, lasting 100 days per year, with a double amount on years four, five, and seven. After the complete cycle, the *gyōja* has clocked up about 29,000 miles! That amounts to a stretch of more than a lap of the Earth at the equator.[46] This all takes place on or around the same mountain, occurring mostly at night. The timing of their runs suggests that the monk is oriented towards the inner experience, not having a view to take in. It is an exercise to fine-tune one's skill of mindfulness and to put the conditions in place for them to achieve these "non-dual" experiences—where mind and body no longer seem separated. For some Buddhists, these experiences are glimpses of reality as it truly is, free from the mundane interpretation of ordinary humanity. Non-duality while running is a "total" experience.

Through the meditative aspect of running, we can connect with something bigger than ourselves with the flow aspect and also connect to our ancient human ancestors, whose running and hunting our own existence depends upon. Through running, we can get close to our ancestors in terms of empathetic experience. And lessening that gap between one person and another seems to correlate with lessening the gap between body and mind, but that's a theme to be explored later. In the previous chapter, I mentioned the relationship with paths created by walkers in the past, and how we, by repeating their steps, have some sort of connection. Here, it's connection in an intense experience that relates us to our past as a species. Physically, it would have been similar for them, chasing down their prey to exhaustion, perhaps over a whole day or even longer. Our bodies are the same as theirs.

Perhaps, then, by running we are being even *more* human than when we walk in terms of connecting to our embodiment. Although *erectus* and *neanderthalensis* also made for excellent runners, *sapiens* were better again, being that bit more efficient. When I posed the question at the start of this chapter as to how it could be that we have the identity of being runners—even including those who may have never have run in their lives—it's because a runner is something we *are*, and not something we do. With running then, we are *being* and *doing* at the same time.

So far, we have just two fundamental biological characteristics of movement that have endured up to contemporary times: walking on two feet and this amazing ability for endurance running. They are the basic locomotory expressions of humanity. And with just these biological tools, a tiny number of curious adventurers would take over the world, and it is to the expression of this where we turn next.

CHAPTER 4

Peeking Beyond the Horizon

We shall not cease from exploration
And the end of all our exploring
Will be to arrive where we started
And know the place for the first time.[1]
—T. S. Eliot

Something rather striking for an extraterrestrial visitor coming to Earth would be the fact that people are *everywhere*. No matter where a Martian, say, may land, it's likely that it would be not far from some type of human community. This is not only the case at the dawn of the twenty-first century, but has been so for many generations. All in all, this single species has today covered roughly half of the land on the planet—mostly on two feet, leaving their produce and waste everywhere. Today, there are previously civilized areas which are now wild, such as the southern Amazon basin.[2]

From the harshest northern climates inhabited by Inuit people, to the brutal deserts of Africa where one finds the Berbers, humans have made themselves pretty much a ubiquitous species. At an altitude of almost 12,000 feet, there are about 2.5 million inhabitants of La Paz, in Bolivia—the highest capital city in the world. In the Australian desert there are the descendants of Greek migrants living in underground tunnels in what was the small mining town of Coober Pedy. By about 15,000 years ago, the Americas were finally conquered, leaving only Antarctica uninhabited until modern times. They got to a point in their story where there were few unexplored territories left, and it's fair to say that humankind considers the Earth to be *theirs*. Through their bipedalism, *Homo sapiens* have achieved a totality of dominance.

They operated at the same level of intelligence that we use today and a single factor of that type of mind was to be the driving force of much of this colonial accomplishment—the faculty of curiosity. It is the intellectual equivalent of the urge to move mentioned in Chapter 1. This innate desire to know is what initiates our dreaming of faraway places and trying out new things.

Curiosity

Humans typically cannot rest until they have explored, upended, poked, and figured out anything that they do not yet comprehend. It isn't just a feature of some individuals but a flagship characteristic of our entire species just like our hairlessness or bipedalism. It must lie at a layer deep within the psyche given that we insist on covering the ground of the unfamiliar until it is finally familiar.[3] In short, we *undo* unfamiliarity.

Curiosity drives everything from childhood inquisitiveness to the scientific method. It compels us to watch movies until the end, climb trees, and read books. It keeps us interested in the lives of our friends and relatives aiding our empathetic capacity to care and to love. Most of all it provides us with a very special yearning: to see what's around the corner, to peek beyond the horizon. We also turn it inwards, in self-curiosity, which often manifests in spiritual or philosophical ways. One of the most famous observations of Socrates is that the unexamined life is not worth living.[4]

Inquisitiveness is mostly about learning and remembering. Humans have an insatiable appetite for filling their enormous heads with information. Although curiosity is an intangible characteristic, it nevertheless informs much of our movement, as we shall see. Human curiosity is not confined to just *Homo sapiens*, nor is it a recent phenomenon. We are not special in this regard. As we saw in Chapter 2, *Homo erectus* learned to control fire, and traveled far and wide.[5] Curiosity therefore is most probably a characteristic that lies very deeply within us. It is *primal.*

Dave Scott, an astronaut on the Apollo 15 mission said of his experience after landing on the moon, "As I stand out here in the wonders of the unknown ... I sort of realize there's a fundamental truth to our nature: man must explore."[6] It was this charge that drove modern humans not only to colonizing the entire world, but reaching the Moon. The most remote place on Earth—Rapa Nui (Easter Island)—has had human civilization for about 1,600 years. We are like what biologist E.O. Wilson calls "extremophiles," the bacteria that thrive in extremely inhospitable environments such as in the upper atmosphere or within glaciers.[7] Our method was nomadism. The first peoples were constantly on the move. For roughly 95 percent of our history as a species this is how we lived. Throughout the world, our ancestors wandered, hunting, gathering fruits and vegetables, and fishing, always on our feet, always on the move.

The first modern *Homo sapiens* in this world found themselves leaving the Great Rift Valley in today's Ethiopia. There were two exit passages, one to the North at Sinai and the other to the East, leaving from near the horn of Africa to southern Arabia. The first exodus of modern humans to leave Africa did so around 120,000 years ago through the North passage, over the Red Sea.[8] They settled, lived, and eventually died out in the Levant—where Syria and Jordan are today. Another exodus of modern humans occurred about 85,000 years ago, this time via the East passage, crossing the Bab-el-Mandeb strait.[9] It was these people who were to evade extinction and populate the planet. Within 10,000 years their descendants had managed to colonize Australia. In the far North, about 16,000 years ago, due to lower sea levels, the landmass known as Beringia provided an intact connection between Eastern Siberia and what we know as Alaska today. The first Americans were born of this crossing and began the slow process of spreading down through the entire length of both continents, which took approximately 12,000 years to achieve.[10]

Up until the 1980s there were two conflicting theories of our origins. The first was the "multi-regional" model, which proposed that the various human populations today are descendants from different prehistoric human populations. This accounted for the regional variations that are often understood to be "races," and that the lineage of some human populations today stemmed from *Homo erectus*, or *Homo neanderthalensis*. There is an implicit suggestion that these ancient populations were much more attached to land, and were probably not colonizers. If it were true, it would mean that humans are naturally not movers in the way I have been describing in the previous chapters.

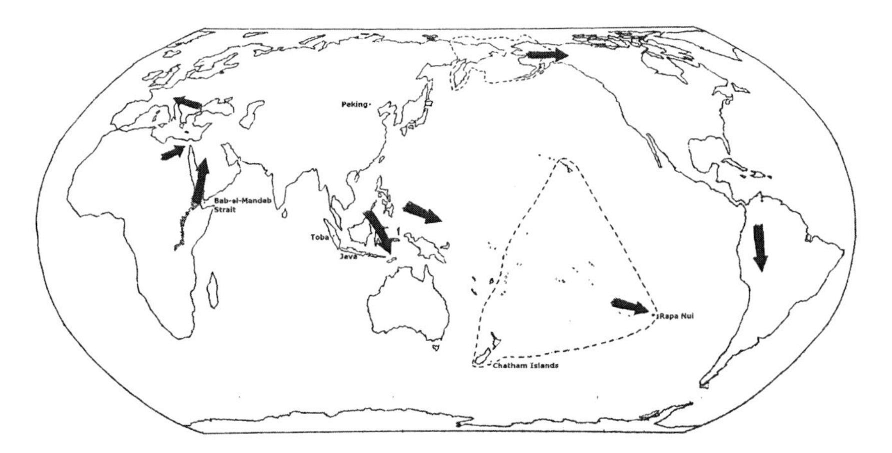

Fig. 5. World map of human expansion. The Great Rift Valley is the shaded area in East Africa. The dotted lines indicate the areas of Beringia and Polynesia.

The competing view was the "recent African origin" model, often shortened to RAO. This claimed that all humans are essentially a single enormous family, which comes ultimately from a single exodus from Africa. This idea was first proposed by none other than Charles Darwin himself.[11] It shows the modern human to be an adventurous hunter-gatherer, who is conditioned by immense curiosity and daring. For a long time, favoring one model over another was merely a matter of opinion. That was until the field of genetics began to blossom.

Adam and Eve

Often, when thinking of genes, one typically considers the DNA of the nucleus, that is in every cell of our bodies, which is inherited from our parents at the moment of conception. There is, however, another set of DNA which lies in the mitochondria, which are the energy producing parts of the cells. Known as mitochondrial DNA (mtDNA), it is of the upmost importance for realizing our genetic history. This is because of two unusual qualities that it has, making it quite different to "nuclear" DNA.

The first characteristic is that mtDNA follows the matrilineal side of inheritance. When sex cells (sperm and egg) are produced, with half the nuclear DNA included in each (twenty-three chromosomes in humans), the sperm uses up the energy in its travels towards the egg. By the time that the sperm fuses with it, there is nothing left other than those twenty-three nuclear chromosomes. Everything else has literally been burned up on its journey to get to that point. Consequently, none of the male mtDNA is included in the fused egg.

Those genes end up in a sort of lucky dip with the corresponding ones from the egg, and new combinations randomly form. This brings us to the second unique characteristic. MtDNA doesn't recombine, because there is nothing for it to recombine with. Mitochondrial genes simply get passed on, as they don't enter into the lottery. They do mutate—as all genes do—but only as the cells replicate. So, mutation within mtDNA happens with a clock-like regularity, meaning that traits related to it are traceable through time. It is in these traits of mitochondrial DNA where the secrets of our origins lie, because it's possible to reverse the process and trace back from daughter to mother in a continuous line.

This is precisely what a team of three biochemists did at the University of California at Berkeley in the mid–1980s, and the result of their work was dumbfounding.[12] Tracing back through an unbroken matrilineal heritage, they found that they could identify a single individual who was essentially an ancestral mother to all people. Amazingly, a "Mitochondrial Eve" once must have existed. More recent studies have located her to have existed in Southern Africa—precisely where the San live today—living about 200,000 years ago, and it is from her mtDNA that the people who have populated the whole world come.[13] Not that this person was the first human, but simply was a single woman to whom all living humans are related today.

This was the proof that the RAO model of human evolution needed. It is now the standard accepted model, thus putting the multi-regional model in the bin. Other lineages died off, like those in the Levant, leaving one out-of-Africa line, known scientifically as the "L3 haplogroup," to populate the non-African world, the other haplogroups remaining in the Southern part of Africa.

In addition to Eve, there is an equivalent male lineage. This was deduced by following the Y chromosome, which is found in nuclear DNA, but only in males. The Y chromosome is what makes men male, so to speak, and is hence traceable in the same way because it passes only from father to son. Just like mitochondrial DNA, the genes on the Y chromosome don't recombine either and are therefore traceable back along the male line. Again, all DNA mutates as it replicates itself, and one of these mutations—known as M168—was found among all male populations outside of Africa.[14] And so we have a "Y-chromosomal Adam," thought to have been a single individual who lived between 120,000 and 150,000 years ago.[15]

When including all living humans today, we are still a species with surprisingly small diversity when it comes to our genetic heritage. To put that in perspective, the world population of humans demonstrates only the same genetic variation as what one would find within a localized group of another animal, living in small numbers. Human diversity at its greatest clocks up about 0.3 percent whereas orangutans, although very few in number, measure about 5 percent.[16] We are an extended family, all of us humans today, despite the superficial differences in appearance. The reason that this may seem contrary to our intuition is because our cultures tend to focus on differences rather than similarities, sometimes creating a false sense of "us and them."

A Bottleneck

This lack of genetic diversity is an anomaly, and begs the question: why so little? When it comes to the narrow band of genetic diversity, the common ancestors, and also how recent our species is compared to *erectus* and *neanderthalensis*, the numbers are simply staggering. The highest estimates of the group from which stemmed the covering of the Earth is only about a thousand individuals, and the lowest is as few as 160![17] *Homo sapiens* seemingly have managed to overcome massive odds stacked against their survival. Regardless of the actual number—probably it will never truly be known—it was astonishingly small, even at the highest estimate. Coming from a population bottleneck, what these people and their descendants achieved was truly remarkable. We're all descendants of a tiny group of humans who laid their legacy through their movement.

If the RAO model of human migration is correct, we would expect there to be more diversity among sub-Saharan Africans today, and that is precisely what was found. In the very same areas where the San continue their prehistoric hunting methods, there is more diversity than in the *rest of the world combined*.[18] And even across the European-Asian continent today, there is far less diversity to be found than between Eurasians and Africans.[19]

The question still remains as to why so few people, and there's not much hard evidence to support any working theory yet. One such idea—known as The Toba Catastrophe Theory—posits that an eruption of a super volcano on the Indonesian island of Sumatra caused a "volcanic winter" 75,000 years ago, its ash dimming the sunlight and contributing to an acid rain phenomenon damaging foliage the world over. The debate still rages over the solidity of the theory, particularly after stone tools were found both above and below the layer of Toba ash in Malawi, suggesting that although the ash made it several thousand miles across the Indian ocean, it didn't cause a mass human extinction.[20] Either way, the genetic evidence still points to there being an infinitesimal population at the time. In today's terminology, *Homo sapiens* were once an endangered species.

But on the Southeast side of Sumatra lies the island of Java, where one of the most important and famous discoveries of prehistoric humans happened. In the early 1890s, when Darwin's idea of our species' origin was still a hotly debated topic, Eugène Dubois, a Dutch paleoanthropologist, discovered the remains of a man, later to be simply called "Java Man." Dubois claimed something highly controversial: that this man was a sort of transitional animal between apes and humans. The reason? Because he hadn't found *Homo sapiens* at all. Java Man is the remains of *Homo erectus*. This other human species made it as far as Indonesia, and the sea levels in that part of the world were not such that a land bridge existed, like in Beringia.[21] This find was backed up by "Peking Man," found in the 1920s in Northern China, and since then about 100 similar remains have been found on Java.[22]

Islands

It seems that for both *sapiens* and *erectus*, walking and running wasn't enough. Nor was simply being near the water. They wanted to traverse it! Their kinetic urge made a leap to operate beyond the body. A wholly new method of movement was invented. It's been suggested that some may have learned through observation of animals floating on driftwood during times of flooding.[23] How people have gotten to islands is a topic which fascinates many, best exemplified by the success of Thor Heyerdahl's wonderful *The Kon Tiki Expedition* (1948)—a daring account of sailing West from Peru to French Polynesia on a prehistoric raft.

Sea travel plays a major part in ancient tales, such as *The Epic of Gilgamesh* and in Homer's *Odyssey*, which provide the very foundations of literature and storytelling. That element of travel—mentioned in Chapter 1—has a deep association with the sea. When humans see a horizon line of the sky meeting the water, it seems to beckon them to peek beyond it. Unlike running or walking, there is a special thrill to using one's body in an inventive manner, steering a craft of human ingenuity and managing the swells and dangers of the sea. Later all can be recounted, complete with mysterious creatures and monsters of the depths. According to some stories, the sea has a version of everything on land, but more macabre and menacing.[24] Indeed, it is because of the types of stories that are told that we know some part of the history of how the farthest reaches of human colonization happened.

Fifty thousand years ago, the colonization of the Pacific began, and continued in spurts, the last one being within the last 3,500 years, probably finishing in the remote Chatham Islands about 1,000 years ago.[25] Over such a broad expanse of space and over a similarly broad expanse of time, it is no surprise that the Polynesian islands demonstrate a diversity similar to that of genes in Southern Africa, but manifest in their palette of languages, which averages out at about one language for every 3,500 people.[26]

Much of why this is so astonishing to us today is because of our cultural view of the sea. We understand it to be a hostile barrier. Consequently, we see islands as pristine petri dishes where human experiments come to pass, just like in William Golding's *Lord of the Flies* (1954). But the more we examine these societies the more we realize that the sea was, in fact, a great provider of access right up until the invention of steam power.[27] Indeed, before steam, things traveled far faster by sea than by land, and at a tiny fraction of the cost. It's easy to overlook this, not simply because our cultures are now so land-focused, but also because there exist so few maritime archaeological finds.

After the migration out of the Great Rift Valley in East Africa, humans finally made a great leap from land migration. They began inhabiting islands, a behavior that has been described as an "act of faith."[28] This is a significant kinetic change, one which was present in embryonic form with the throwing of spears, and which would repeat itself throughout history henceforth; namely, the invention of new methods of movement, extending our kinesis beyond our bodies. Harnessing a force that lay in their environment, they managed to take advantage of the wind. The technological invention that extended swimming or drifting is a remarkable achievement and it would happen with the other locomotory modes of kinesis which are the subjects of chapters to come. The inhabitation of islands represents humans at the height of their powers; expressing curiosity, self-challenge, development of knowledge, exploration, and ultimately trust that things will actually work out. These dangerous undertakings are proof of the extraordinary blend of natures in the human being.

And dangerous it certainly was. The last parts of the world to be settled by humans were the small islands of the Pacific Ocean such as Polynesia, lying at immense distances from any mainland. And those who managed it needed a hopeful vision as a counterpart to the cultural change of leaving behind the nomadic life.

On the open sea—meaning where one has lost sight of land—their understanding of space is recorded to have been conceptually very unusual. Given that there is only sky and water visible if navigating in daytime, a strange stillness becomes part of the kinetic experience. Understood as if the boat was still, the islands came and went from them. The world around the boat was in motion, in a smooth space, one not measured or marked, one where these things are impossible, a field of softness, constant change and complete lack of fixity. They sailed in dugout canoes, paying attention to breezes and smells on the wind, the presence or absence of turtles and seaweed, and to the color of the water.[29] Some are reported even to be able to smell the reefs beneath the water's surface, and see a phosphorescence some yards below when nearing land.[30] The knowledge employed was such that never was there a situation that was unknown to the navigator.[31] For this first time, human kinetic effort was not autotelic, their effort was a planned investment—the boat being a fabricated item specifically for that purpose—and once afloat, the wind took them with a new speed and that investment paid off.

In tandem with their highly developed navigation skills, the Pacific sailors also invented the outrigger canoe and the "crab claw" sail.[32] This is a triangular rig, far easier to use than the European rectangular rigs, and allowed them to navigate *into* the wind, tacking easily. Over the years of research into the peoples of the Pacific, many Eurocentric theories have suggested that they were populated by simply people drifting on rafts, and after many failed attempts some became lucky. However, data (and attitudes) are now better and it became clear after some researchers did computer simulations of similar journeys following trade winds, that this exploration and populating must have been intentional.[33] The settlers also took enough people, plants, and animals to create new island societies.

Ship Burials

Sometime before the last of the eastern Polynesian islands were reached by the first humans, there was an intimate connection between people and their ships forming elsewhere. Far away, on the other side of the world, the Egyptians developed a culture of ship burials. Known as Solar Ships, they were buried as part of their funerary rites of their owners. The best-known example dates back some 5,000 years, and this—called the Khufu ship—has never even been in the water. Its sole function was ritualistic. It was found in its own grave, beneath the Great Pyramid at Giza, deconstructed in orderly fashion in fifteen separate burial layers.[34] This was no token item either, measuring a whopping 140 feet long and twenty wide.[35] Many other examples have been found, suggesting that it was a somewhat common practice for high-ranking individuals such as Pharaohs.[36] This extraordinary practice lasted for thousands of years, at least until Pharaoh Senusret III, who had one at his funeral in 1846 BCE.

What they represented specifically we may never know, but it is certain that they were important considering their views on death and the afterlife. Even though the Egyptian society at the time had yet to discover the wheel, it was nevertheless one of the most complex civilizations in the world. Highly religious and ordered, this culture must have valued the vehicular nature of passing to the afterlife, kinesis extending into non-physical realms that we see as still and static.

Whatever it is, there is a remarkable connection with the outward expression of human movement, via a complex-built instrument to enable that movement, in perhaps the most significant spiritual context there is, one's death—where the body's movement ceases entirely. Being laid to rest in a vehicle must be the ultimate expression of the marriage of kinesis and stasis. In death, the repose of total stillness, carried forever beneath the ground in a ship, one moves in another sense.

What's interesting about this cultural activity is that these Egyptian ship burials were the first of many examples of such activity. Again, in a culture with ostensibly no connection to the Egyptians, the Vikings also buried ships. But here the difference was that the deceased were actually buried *within* the ships. Only four such examples have been found, the best preserved being the wonderful Oseberg Ship, constructed in 834 CE.[37] Discovered in a large burial mound in Norway in 1904, this is a large sea-going craft, sixty-five feet long. The bodies of two women were found in the ship, along with sleighs, a cart, all very elaborately carved and decorated, and fifteen horses. Today it remains as one of the best Viking discoveries, large enough to need thirty rowers to pull, with the typical dramatic rising bow and stern all richly embellished. It suggests a connection of the journeying person after their last rite of passage, where one can no longer move, but is instead moved by a ship.

Interestingly, many modern Christian churches are structurally reminiscent of boats. A decommissioned ship is the epitome of "post-movement." The sea gives us a subtle sense of connection with that inner desire for stillness, which may explain why we're drawn to water for reasons seemingly beyond natural beauty, to enjoy deeper relaxation and other benefits.[38]

The extension of human kinesis was really a remarkable achievement. For the first time in our history, we were moving without depending upon our bipedalism. Our motion became externalized. In so doing, we created a new sense of space, one that was wholly new, a container freely drifting through the infinity that had laid before us. As Michel Foucault muses, "the boat is a floating piece of space, a place without a place, that exists by itself, that is closed in on itself and at the same time is given over to the infinity of the sea."[39]

Migration

Many complex events were to unfold by the power of this extension of movement. Through the ship, the shape of today's world—in terms of economic development and trade—was set up. The explosion of trade in the mid-sixteenth century and colonial expansion from then can be counted as among the principle historical forces for the way the world works today. Foucault continues, "... it goes as far as the colonies in search of the most precious treasures ... the boat [has been] the great instrument for economic development.... In civilizations without boats, dreams dry up, espionage takes the place of adventure, and the police take the place of pirates."[40]

Boats accelerated the humanizing of the Earth, which in turn defined us as a species. Migration historian Peter Bellwood tells us that "Without migration there would be no human species, at least not outside a small region of Africa."[41] As nomads, we essentially adapted to migrate, to colonize. Our behavior of migration changed in tandem with our physical evolution, a sort of co-evolution between culture and genes.[42] And in turn, our existence today *depends* upon those migrations as a historical force.

Unfortunately, nowadays there exist strong myths of belonging to particular territories. We have the amazing technology to carry out DNA tests but we are misleadingly informed that we are a certain percentage "Italian," "Brazilian," or "Korean," as if these are biologically relevant entities. It is an absurdity to think that there is anything genetic about "being French," say. What could be inherently French in terms of human biology? Even without an evolutionary bottleneck, there are no genes from specific countries. Genes are simply inherited codes for the traits that may or may not be in the DNA of a given individual. They—just like us—don't naturally belong anywhere. The overwhelming majority of human history is nomadic and migratory anyway. Its prevalence means that there is no such thing as migration as a particular expression of motion. Rather, it is a generic expression of it; *migration is being human.*

It is predicted that migration will be once again a typical form of human living,[43] and there are more migrants today than ever before in world history—given that the prehistoric population was so small. One billion people are considered as migrants today.[44] Even the 2016 Olympics in Rio de Janeiro hosted a refugee Olympic team, something we will probably see commonly in the future. Modern migration is a different matter from the nomadic colonization of the world that I have described. It is different because settled lives didn't yet exist and nomadism was an inherent quality of their being. One philosopher of movement and migration describes today's migrants perfectly succinctly: "The migrant is the political figure of movement."[45] What constitutes modern migration is "social motion" and is inherently political.

Our prehistoric ancestors' nomadism, curiosity, and colonization all intertwine, exemplifying a different mindset from ours entirely, and took place in societies before such highly ordered civilizations arose. They had the capacity for an untold amount of knowledge, augmented by their capacity to communicate.[46] They were to dominate the Earth in every manner over the next 50,000 years, acting directly upon their kinetic urges. Our early ancestors seem to have had a consciousness that brought together both the mental curiosity to see how the world works, and the eagerness to physically discover it.

What was all this for, we may ask? Whatever the intentions, it could be read as an expression of the kinetic urge. But what came after was the greatest developmental shift imaginable, an expression of the static urge.

The first settled societies came about and germinated the sedentary lives that so many of us live today. They were to continue extending their motion like their ancestors' sailing feats, increasingly through the application of mind more than body, relying on kinetic invention. And not only new forms of movement would arise from this, but ultimately new forms of space, consciousness, and mind.

Chapter 5

Abandoning Nomadism

Where Do We Come From? What Are We? Where Are We Going?[1]

—Paul Gauguin

By the time the processes of migration and colonization began to slow down, human motion was poised to enter into a new phase. *Homo sapiens* gradually went from being an endemic species—meaning located to a specific area—to one of cosmopolitan distribution. In other words, their presence got close to a saturation point. Curiosity had brought them traipsing over every corner of the world with very few exceptions (Antarctica and Madagascar were peopled a good deal later).[2] By 15,000 years ago the southern part of the Americas was populated, meaning that the human takeover was coming to its peak.

This chapter is about what happens next in mankind's story of movement, where the focus shifts a little. The tension is still there between the kinetic urge—to move and explore; and the static urge—to be still, to transcend motion. But now the static urge comes to the fore. Over a period of about six thousand years—a short time considering the scale with which we have been dealing—we stopped moving in the way we had been doing before. Without doubt, this is the greatest single change in humanity's history. It has had major consequences which we are still feeling today. Almost all human cultures follow this pattern of living a settled life, having completely abandoned nomadism, leaving only a few exceptions.

Traditionally, this is the point when prehistory ends and history begins. Both are in the past of course, but the latter is where people began to write history. The moment where history begins is also the point at which an understanding of time seemingly becomes important. The early settled civilizations demonstrated this importance of time and the past, as can be seen from some of the fragments that survive from their writings. Time itself was seen by some as simply a measure of motion.[3] This will be of importance later, as it shows that these cultures were living somewhat *outside* of themselves. People such as the Sumerians wrote their own histories, relating to time and therefore themselves in a new way. It was as if they'd acquired a new self-awareness and could look in from outside. They began relating to space in a new way too, the natural environment falling under their control. It was a big change and later played an important part in the story of kinesis. But for now, back to those first settlers.

The Epipaleolithic

Until quite recently, it was thought that the settling of nomads was a consequence of farming. The historical model was that the Agricultural Revolution (also called the Neolithic Revolution) preceded both long-term settled living and the ensuing population explosion. The produce of agriculture replaced the gleaned food that had previously come from hunting, gathering, and fishing. This was thought to be because of dramatic changes in the weather at the time—a return to glacial conditions during a cold period known as the *Younger Dryas*.[4] Long-distance motion was therefore no longer necessary, and people began to move less.

But there was a problem with this hypothesis: what agriculture needs is a relatively stable, settled society *already in place*. Without this, early farmers could not have experimented with their crops because each trial takes a full year to see through. Unlike political or technological revolutions, there was far more slippage in time with the development of agriculture, as there was much starting and stopping, retrogressing and starting anew. Just like with the hunting societies before them, information could be shared quickly throughout and between cultures; but the information itself was slow to be applied or develop.

In the mid-twentieth century in central Turkey, archaeological sites were discovered which turned the tables on our understanding of the order of prehistorical events. It is here that many of the first human constructions survive. In archaeological jargon, these were *epipaleolithic* cultures, meaning that on the timeline, they belong sandwiched between the nomadic hunter-gatherers and the first domestication of wheat and barley. The archaeological evidence suggests that it had previously been understood the wrong way around. These settlements demonstrate urban societies existing *before* any evidence of agricultural activity. Writing about this, the experts were definitive. "We are now aware ... that sedentary village life began several millennia before the end of the late glacial period, and the full-scale adoption of agriculture and stock rearing occurred much later, in the late ninth and eighth millennia BC," say archaeologists Peter Akkermans and Glenn Schwartz. "It is now evident," they continue, "that agriculture was not a necessary prerequisite for sedentary life, nor were sedentary settlers always farmers."[5] The first agriculturalists had pre-adapted to this form of living by living in settled hunter-gatherer communities, and one culture in particular—known as the Natufian—continued in hunter-gatherer mode throughout the Younger Dryas period. All this points to the current thinking that weather was not a contributing factor to the birth of agriculture.[6] For some, it would seem that sedentism was not merely at the cusp of agriculture, but looks as if it preceded it by some thousands of years.[7]

Settle Down First

Cities sprang up in various parts of the world and agriculture followed sometime after. This happened *polygenetically*—meaning that many cultures arose independently of each other. There were many centers of urbanism worldwide: Mesopotamia, India, China, Central and South America; with farming also beginning in New Guinea[8] and Ethiopia.[9] Of course, all these cities didn't come into existence at precisely the same time, but across a band from about 9,000 years ago to the later cities in the Americas about 4,500 years ago. Some cities in the Americas (Caral in Peru, for example) show that their people lasted as long as 1,000 years without engaging in any war.[10]

It's logical to assume the need for order to be in place for social cohesion to occur. Every society functions along a set of cultural "norms," rules often implemented by a hierarchical power structure, be it religious, ideological, or legal. The smaller, hunter-gatherer societies were highly egalitarian compared with those of the epipaleolithic and neolithic.[11] Even today's hunter-gatherer societies, such as the persistence-hunting San whom we met in Chapter 3, are "aggressively egalitarian," to borrow a phrase from an archaeologist working at the forefront of this research.[12] So, the city needed something to exist, to be commonly believed in, and put in place as an overruling power so that the society would function. It was a system of beliefs, rather than one of food production, that allowed for the change from nomadic to settled ways of life. Responding to the question of how these far bigger societies were to function, best-selling historian Yuval Noah Harari tells us that "The short answer is that humans created imagined orders and devised scripts. These two inventions filled the gaps left by our biological inheritance."[13]

The imagined orders he refers to are the new religious and social systems that came about at this time, evidenced in the remains of these ancient cities. This is because the larger the size of a given community, the more a system of rules, manners, etiquettes, and commonly held beliefs is needed. This is why ethical laws have an invariable tendency to come from authoritarian superhuman forces, such as the ten commandments given to Moses by God rather than being devised by Moses himself; and the general upholding of scripture as a guide to ethical instruction in many other religions. And most ironically, despite a new ethical order, there was a huge drop in the level of egalitarianism in the new way of life. Indeed, many of today's problems of inequality can be traced back to ownership of food-producing land.[14] Such is the setback of a non-nomadic culture.

Spirituality and The City

The settlement of Göbekli Tepe in Turkey contains what are now thought to be the oldest human-made structures in history.[15] About twenty huge T-shaped carved monoliths, reaching to about twenty-three feet in height and forming a circular structure were found here. When underground, they were still large enough for the tops to appear to be medieval gravestones to those who first examined the site in 1963.[16] That archaeological dig revealed that around the circle were a series of dwellings which had been constructed some time later. Since then, the picture has become clearer: it was spirituality that led to the first building projects on Earth.[17] More recently, studies revealed that one of these T-stones contained the oldest example of a written word. What word was it? "God."[18]

Göbekli Tepe began as a center of religious worship where nomadic people came specifically for that purpose.[19] Another small city at Çatalhöyük had a similar stone circle at its center, and was also visited regularly by hunter-gatherers, later seeing settled dwellings too.[20] Çatalhöyük in particular is interesting because the society there lasted so long—almost 2,000 years. The existence of a stable society for this long is pretty astonishing, given that it followed such a radical change to the way of life. And this was precisely the set of conditions for agricultural experimentation to occur, giving rise to domestications of wheat, barley, and sheep. Like the mysterious ship burials in the last chapter, it's possible that these places were centers of funerary cults, and once agriculture was developed, they became fertility cults.[21]

All over the world, humans in various societies began to live under similar static, settled conditions. And if not caused by weather, they certainly became dependent upon it. Their food was now less secure than before, begging the question as to why this occurred in the first place. Settled people were more vulnerable to disease, malnutrition, and crop

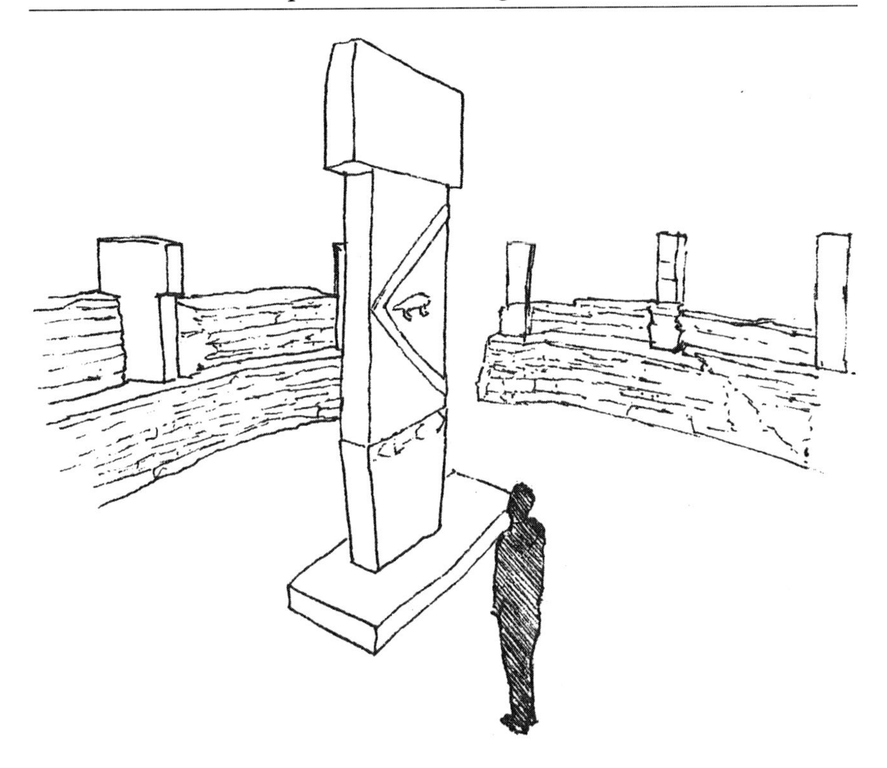

Fig. 6. The T-shaped pillar no. 18 at Göbekli Tepe, in an artist's depiction.

failures with quality-of-life suffering tremendously during this change, leading one popular historian to label the Neolithic Revolution as the "worst mistake in human history."[22] It does seem logical that a society dependent on certain weather patterns for their food would worship the mysterious gods of rain and sunshine. And although details of how this all came about may be a little unclear, two things stand out: that spirituality was deeply connected with the settling of our ancestors; and that this can be seen to represent the static urge I described in Chapter 1.

In a relatively short period of time, the nomads went from bands of about 100 people to large societies. In Çatalhöyük for example, there was a population of about 10,000 people.[23] It was a high-density situation which used a spirituality-derived order to keep things functioning. No longer being able to simply up and leave if there was a disagreement, there was now a more complex set of consequences to actions and words, and so a system of ethics was necessary. Before agriculture, there was a simultaneous change in how people expressed themselves kinetically (by settling) and how they expressed themselves spiritually (by building these temples).

After early agriculture took hold, high-density places of habitation arose and quickly became sophisticated areas to live in. The Sumerian city of Ur was one of the first big cities in history, and what remains of it includes fireplaces in every house. It demonstrates a knowledge economy, with school classrooms and clay tablets of multiplication tables. Not only were they taught arithmetic, but one of the best-known exercises of mental abstraction of space—the Pythagorean Theorem—made its first appearance here, some 1,500 years before the great Greek mathematician and philosopher discovered it for himself.[24] Not only were they sophisticated places, they held surprisingly large populations. Underground storage silos were found beneath another Sumerian city—Shuruppak—which were large enough to store food for 20,000 people.[25] Ur itself was home to more than 60,000 people, being the first known urban center.[26] This is all 6,000 years ago, a time when a massive proportion of the Mesopotamian population were living in cities—about 70 percent.[27] These urban centers, therefore, were not freak accidents, but part of a "land of cities."[28]

The city is the dialectic of agriculture. As a pair of interlocking opposites, each needs the other to survive: the farmlands need the city in order for there to be a marketplace to sell its produce; and the inhabitants of the city need the food that cannot be easily grown in their urban environment. Despite being cultural opposites in a way, they nevertheless depend on each other: one to provide the food, the other to pay for it. The city and the surrounding farmland co-exist in duality by necessity. As the urban population increases, more land must be used to produce their food, resulting in a feedback loop. Unfortunately, this constant expansion is unsustainable.

Entanglement

The Epipaleolithic period and the Neolithic that followed it combined to form a special epoch in the story of human motion. This was the moment where each of our conflicting urges manifested themselves outside the body. We can call these entanglement and extension, and I'll deal with each in turn. The first is based on the urge for stasis, or the urge I described as that of transcendence of motion, to achieve a total rest, going beyond desires altogether. It is a yearning that operates more on a spiritual plane, being less intellectually or emotionally based. Perhaps we could say that it's a desire to overcome desire itself. We could also understand it to be associated with a certain "calling" to where there is no change. Given that a spiritual shift was central to the conditions of the first settlers, this is important.

The way this urge manifested itself was through our sudden connection with what some experts simply call "things." A thing is just that—an object. Compared to the nomads, settled people owned far more things, and produced more things. The complexity of civilization is based on a huge number of connections between humans and things, all related in a vast web.

Things are the unmoving objects that abstractly relate to or embody our urge to be still. The settlers created a sedentary world of objects, a material world. The concept of ownership comes from these times, which is likely the beginning of our strong attachment to things. Written laws came about as a way to manage property, and trade of things gradually grew to be a huge part of the economy of these societies. The kinetic was made static; and the static was objectified and embodied in things.

We might look back to the ideas of Heraclitus and Cratylus here. As discussed in Chapter 1, the idea of Heraclitus was that the only reality is motion. Cratylus took it a step further and said that because the only reality is motion, there is no real thingness at all. Nothing is fixed, in his view, so nothing has any real substance. Now we don't really know what the nomadic bands or early farmers were thinking about this—or if they had these types of thoughts at all. But our concept of this thingness dates from those early settled civilizations. One of the costs of living the settled life is that we deny the reality that both Heraclitus and Cratylus are pointing out to us. We resist the idea that things are in motion by their inherent nature; and we see things as things-in-themselves, having their own "being," so to speak. An oar isn't simply an oar to row with. When it's not being used, it's still an oar. It doesn't become any more "oarish" when being used, or any less "oarish" when used as a lever to lift a heavy stone. But the tendency is to think of it as having a sort of inherent essential thingness regardless. Same with a millstone, a spoon, a bicycle, or a vase. By thinking always in terms of things, we reduce their kinetic nature.

And as time went on, the web of interrelations between humans and things grew enormously. As Ian Hodder, the lead archaeologist of Çatalhöyük, puts it, we became "thing entangled."[29] The production of food, pottery, and other goods increased, and so did the strength of sedentism.[30] Materialism—the overvaluing and fetishism of material things—was more than a side-effect, it was genuinely at the core of this transformation, probably starting with domesticated plants and pottery. Hodder suggests that it was not entirely volitional by saying that "humans get trapped looking after wheats that will not reproduce without them."[31]

If we trace this relationship of thing entanglement to today's secular, hyper-consumeristic societies, we can very easily see the desire for material things, how it's never satisfied, and is probably rooted in some deeper desire left unquenched. There seems to be a loosely inverse relationship between the complexity of a given civilization and sedentism, which lasts today. By complexity, I mean orders of power, technology, materialism, and a high level of thing entanglement. Complexity, while providing increases in some areas of quality of life, also makes people's lives more dependent. All complex societies must eventually collapse as their growth only ever gives a *diminishing return*, meaning that they are inevitably unsustainable. And this is precisely what finished off many of these societies during the Bronze Age Collapse of about 1170 BCE.[32] Some thousands of years later, the greatest empire in human history—Rome—also collapsed for the same reason: the increasing complexity of maintaining such a huge empire, and costly land-grabbing wars to secure space to grow food.[33] Even though agriculture came sometime after the first settlements, it is still what gave us the modern and complex city.

But at the same time, complexity is natural. Organisms have grown and evolved with increasing complexity over time. Darwin, towards the end of *The Origin of Species* talks in this way about entanglement:

> It is interesting to contemplate an entangled bank, clothed with many plants of many kinds, with birds singing on the bushes, with various insects flitting about, and with worms crawling through the damp earth, and to reflect that these elaborately constructed forms, so different from each other, and dependent on each other in so complex a manner, have all been produced by laws acting around us.[34]

In our world, the many human-thing relationships form to make complex structures which can be seen in many mobility systems such as the roads of the Roman empire, Mongolian postal systems dependent on horses, bicycles in twentieth-century China, and the ships and sea-routes of the Atlantic.[35] All are different in their nature, complex and interdependent, just like the creatures in Darwin's description. Entanglement will be an important element of our story of kinesis.

Extension

Contrary to finding thingness around us, the other extension of ourselves was of the kinetic rather than static type. This was the extension of motion to the realm outside of the human body, much like the boats of early sailors discussed in Chapter 4.

Although the first farmers were settled, there nevertheless remained an itch to explore. This had always been there, and is still within us today. Inner primal curiosity wasn't simply going to go away now that nomadism had been given up. It still remained an inherent part of the human condition. This kinetic desire is the connection point between movement, exploration, and the mind. It began to manifest itself in the most unpredictable of ways, channeling itself into new forms of creativity and leading the Sumerians to develop advanced geometry, mathematics, and literature such as *The Epic of Gilgamesh*. Their sexagesimal system (where the base is six rather than our decimal ten) is still with us today, which is why we have sixty minutes to the hour, and 360 degrees in a circle.

Cities meant that the era of bipedalism being an essential characteristic of the nomadic human was over. Civilization provided us with a feeling of belonging to a place, but we would use new methods to get about. As the lines of travel turned inward, folding in on the urban towns and cities, our kinesis moved outwards from our bodies. We created extensions of motion.

Perhaps for traditional history, it is the rise of agriculture which is the defining change from nomadic tribes to civilization, but for this story it is the extension of human biomechanics into vehicles. In the last chapter we discussed the invention and use of boats for purposes of colonizing islands. But for the "landlubbers," other vehicles would bring about this change. So, with sedentism came new outlets for movement, extending itself as if through a vent by those who were born to move, but were living a settled, agricultural life.

Human kinetic expression was no longer to be that of our physical motion, it would instead be an "extension of man," as Marshall McLuhan famously described communication.[36] These extensions of motion ultimately gave rise to communication between settled communities, as they enabled a fluidity of information. In short, movement *became* communication. And nowhere is this more evident than in the wheel.

A Perfect Circle

The humble wheel was invented somewhere in the ancient Sumerian world, around the third millennium BCE.[37] Oddly enough, it existed firstly as a potter's wheel, before being used as some species of primal rolling device, probably first as a series of logs. The Sumerian culture was one of great technological inventiveness, and as well as coming up with the potter's wheel and moving it to a wagon, they were responsible for cuneiform writing, irrigation, and the lateen sail. But it was in architecture where their forte was: vaulted ceilings, domes, and the arch—the last an invention not rediscovered until the Romans did, the ancient Greeks never demonstrating knowledge of it.[38] Architecture aside, their kinetic expression was of such a level that the remains of leather tires for chariot wheels were found in the city of Ur, as if prophesizing the pneumatic tire of the late nineteenth century.[39]

All under a deeply religious, theocratic Sumerian social order, the potter's wheel was converted into a wagon wheel and later a chariot wheel. Millstones allowed a carbohydrate-rich diet, and within another thousand years the spoked wheel—an item that can carry more than 100 times its own weight[40]—brought us into an era of extended locomotory kinesis in every way.[41] For *the wheel is the basis of modern kinesis.*

Above their heads were their gods, who were worshipped through the structure of the society, and who provided food in return. They were gods of the seasons, sun, rain, water, and the plough.[42] Their subjects diligently observed the motion of the celestial bodies, as evidenced by the astrolabes which have survived, bringing their wheels to reflect the form of the great wheels of the cosmos.[43] The stunning night sky provided a map for sailors at night time, even if the Sumerians were river and coastal sailors, not sailing on the open sea like the Polynesians.

The worship of the Earth and its weather lasted in some form right up until the era of monotheism, which began some two thousand years later. Even today, it's not hard to see the wheel as the greatest invention in history, but it also functions as the link between religion and motion, between spirituality and kinesis. We associate the wheel with time, change, and with life itself. It metaphorizes life, with its cycles of death and regeneration. We could understand it as taking the place of humanity's nomadic expression. The wheel by itself is only motion—pure motion. It doesn't do anything else. Other vehicles are divisible into their parts. But that divisibility stops at the wheel. Functionally speaking, one cannot divide a wheel without making it into another thing. It is the atomic unit of kinesis.

Lewis Mumford, the great American historian and philosopher, once observed that circular motion does not occur very often in nature and that "man himself, in occasional dances and handsprings, is the chief exponent of rotary motion."[44] And it is often said that nature does not provide us with clean, straight lines, or edges. While this is true, it is also true that these straight lines *can* be found once movement is incorporated into the equation. Objects often fall in straight lines. Moving objects demonstrate pure forms, for example say a parachuter falling from a moving airplane will fall in a perfect parabola. Weather systems represent spirals, as do galaxies. Kinesis brings these abstract mathematical patterns to a visible reality. Mumford is correct as far as small-scale earthly nature goes, but in the sky we see many wheels, and heavenly clocks are part of nature too. It might not be so much of a coincidence that the same highly architectural culture who studied and worshipped the heavens, invented the wheel.

Over thousands of years, the Sumerian astrolabe would explode in complexity but maintain its clock-like nature. Planetary models—called *orreries*—were designed not only to represent the heavens, but also to predict the whereabouts of certain planets. They are three-dimensional representations of the systems of the heavens, the oldest dating to about 150 BCE, which could even predict eclipses years in advance.[45]

Unifying astronomy, mathematics, and kinesis, the unceasing line of the circle mirrored the realm of the gods and became the geometrical symbol of perfection. At its heart is not merely its geometric center but the mystery of the ratio between circumference and diameter, π (pi). It is a constant (never changing regardless of the size of the circle), yet irrational number (having an infinite number of digits after the decimal point with no discernable pattern). Furthermore, it is sometimes described as "transcendental," meaning that it comes from somewhere beyond mathematics itself; it cannot be calculated by ordinary mathematical functions (addition, division, square root etc.).[46] Its lure throughout human history, since the Babylonians and Egyptians who discovered it, is revered in awe by many up to and including today.[47]

The perfection of the circle is beautifully described in an anecdote from Renaissance Italy. The story goes that Pope Benedict IX had sent his courtiers to ascertain if the artist Giotto was of a sufficient ability to be brought to Rome for work. Giotto, at their request to do a small sketch, demonstrated his genius by quickly achieving the supposedly impossible feat of drawing a perfect circle freehand. "Here's your drawing ... it's more than sufficient," he told them, and subsequently was called to Rome by the Pope.[48] Far away from Italy, in the Zen Buddhist tradition of Japan, the *enzō* ("circular form") resembles both enlightenment and the universe as a whole. Both of these instances highlight the inherent beauty of the circle that arrests the human being, and how it aesthetically suggests a transcendent nature.

All this comes together to give the wheel a meaning. In equal parts created and discovered by humans, it is the result of settling of nomads, an attempt at perfection, and a symbol of the heavens. It connects the religions founded at places like Göbekli Tepe with the motion that would come from cities like Ur, Uruk, and Babylon. It would catapult humanity forward in countless areas of technology, right up to where we stand today.

Perpetuum Mobile

Of all the wheels in the human imagination, none exemplifies kinetic desire and entanglement more than a *perpetuum mobile*, a self-driving machine which never stops. The earliest and simplest known example of this comes from India in the twelfth century CE, where weights act as levers upon a spinning wheel, continually driving it. It's known as Bhāskhara's wheel or the overbalanced wheel. The problem, of course, is that it doesn't work. Today we know from the first law of thermodynamics that energy cannot be created or destroyed. But what perpetuum mobiles do successfully is show the entanglement between humans, mechanics, curiosity, and desire to overcome the strictures of motion as we know it. Examples of other designs include elaborate water pumps and even a "self-blowing windmill"![49] This shows the desperation of some to control, own, and produce kinesis from nothing, not unlike the tradition of alchemy.

The existential nature of Gauguin's questioning with which I headed this chapter comes from a title of one of his paintings. It's a pithy reflection upon the nature of humanity—at least that's one way to interpret it. I would like to add one more question to his three: *And how are we going to get there?* The myriad of possible answers will be the subjects of

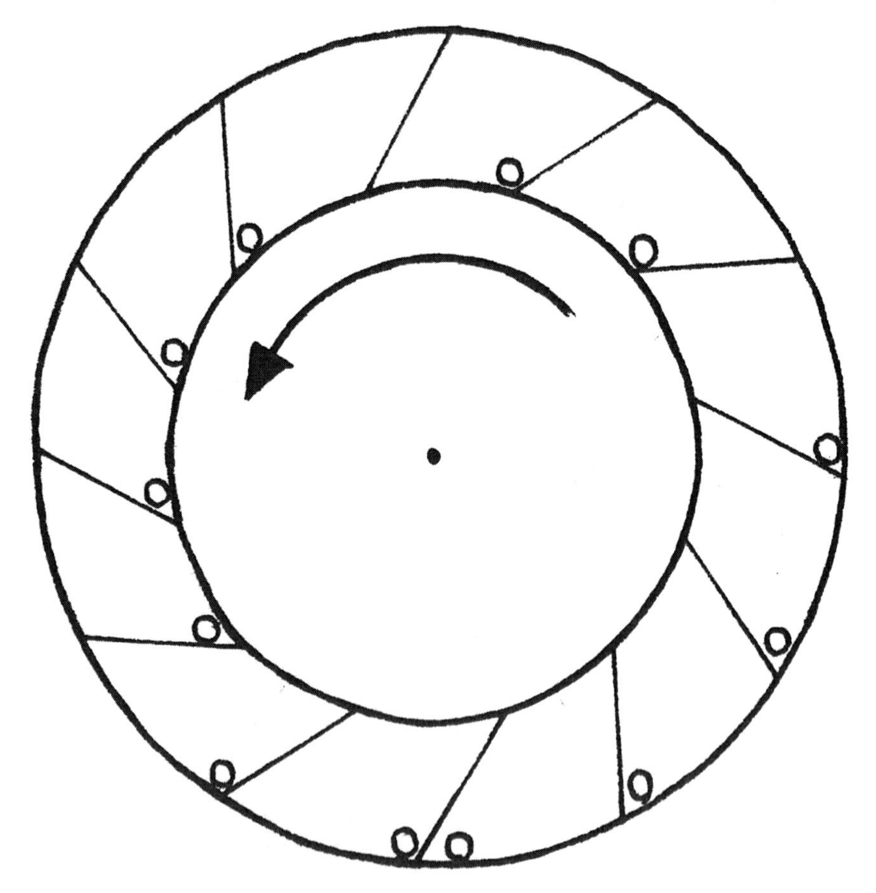

Fig. 7. Bhāskhara's wheel.

the following few chapters. (Incidentally, it's interesting that Gauguin painted this in Tahiti, one of the last islands in the Pacific to be settled.) The first answer to my additional question actually begins with the only non-mechanical method of kinetic extension, an organic moving partner who accompanied us for most of our civilized history. They are one of very few other species to spread across the planet as we did, a companion so deeply embedded in human history that it is said that we are symbiotic, but that remains the theme of the next chapter.

Chapter 6

Six Legs Better

It is easy to conquer the world from the back of a horse.[1]

—Genghis Khan

On a cool December afternoon in 1994, three French cavers were on a day out. Following a walking trail, they entered a well-known small cave. But this time they noticed something unusual, at the back there was a draft. Through a tiny aperture in the cave wall, air was passing through from elsewhere. Despite the falling light, the three scrambled at the rock and removed what had seemingly been a solid wall but was actually a pile of rocks, revealing a narrow passage. Entering it made them the first people to be in that space in tens of thousands of years.

This is how the Chauvet cave was discovered, by Éliette Brunel, Christian Hillaire, and Jean-Marie Chauvet (after whom it was named). A find like this must be every caver's dream, complete with animal bones, stalactites, calcium formations, and vast chambers. Before the government decided to close it off to protect it, they allowed filmmaker Werner Herzog to film his *Cave of Forgotten Dreams* (2010) there. What they were protecting was the cave's magical paintings, now known as the oldest artworks made by *Homo sapiens*. They are as old as their first presence in Europe—going back about 32,000 years, a time when Neanderthals still roamed the continent.

The inner chambers are littered with paintings of bison, bulls, and horses among other animals. Unlike much cave art which can often appear simplistic to modern eyes, these animals are modern, shaded, and textured, sometimes using the natural formations of the rock to give them a three-dimensional aspect. The San rock art includes many animals being chased for food, and indeed there are *zoomorphic* (having animalistic qualities) images present on the T-stones at Göbekli Tepe. But what makes the imagery at Chauvet extraordinary is the inclusion of perspective and motion. An increasing number of experts now believe that the superimposition of images in this and other cave art found in France is an attempt

Fig. 8. Horses depicted in Chauvet cave (above), and a moving image from the cave at La Marche (below), France.

to depict movement. Animals are shown in various aspects of their strides along friezes, and also eight-legged bison whose legs may have moved convincingly against the flickering light from a fire.[2]

One of the most repeated motifs in this prehistoric art is the horse. What is remarkable about it is that they are depicted not just simply being by themselves, but in their *relation* to humans.[3] In the older paintings, such as those of the Chauvet cave, the horses are onlookers, as if observing what's going on around them. They're depicted as sentient, intelligent creatures, but without ignoring their power, elegance, and beauty. Later depictions show them as working animals, in caves as far away as Indonesia.[4] The horse in human art tells us the story of their metamorphosis from wild beasts being hunted for their meat, through being a working animal, to powering early forms of wagons and chariots, and lastly being ridden by humans sitting atop. Not long after the earliest settlements, the horse became fully integrated into human life.

Dawn Horses

The Chauvet paintings demonstrate how far back mankind's fascination with the horse goes. But these creatures were roaming the Earth long before we arrived. The beautifully named "dawn horses" date back to about fifty-five million years ago. To put this in perspective, the last common ancestor between humans and chimps dates back to somewhere between six and seven million years ago.[5] Horses are perhaps the only other large animal species aside from humans that have such incredible abilities to adapt. They can thrive in harsh climates, and fossils have been found of early horse species that demonstrate that they too spread to cover almost the whole Earth.[6]

It's generally well-known that the Spanish Conquistadors imported the horse to the Americas, Hernán Cortés bringing them to the North American mainland in 1519. But fossil evidence shows the horse originally to be an American animal which spread outwards from there. Like humans, horses came very close to extinction but following some unknown miracle, they pulled through and survived to become one of the world's most successful animals in terms of numbers. About twelve million years ago a small number of a single species of the *Hipparion* genus crossed Beringia and colonized Asia and Europe.[7] This is the same passage that the human migrant wanderers would take—as discussed in Chapter 4—but traveling in the opposite direction. Just like the descendants of Mitochondrial Eve and Y-Chromosomal Adam, horses have a very narrow genetic diversity due to having survived a population bottleneck.

Both ourselves and our equine friends are *cursorial* animals, meaning that our limbs have evolved for running, and this is a paramount feature of our bodies. The big difference here is that horses are built for speed. It is their great defense mechanism, being such gently spirited creatures. But over distance they are limited, and in this way a human can in fact almost outrun a horse. This counterintuitive fact is celebrated each year in Wales, at the "Man versus Horse Marathon," a twenty-two-mile race held each June. The event has been won by horses more frequently than humans, but often it's been a difference of no more than a few seconds.[8]

Humans and horses lived in partnership rather than one dominating and enslaving the other and both parties benefited from this alliance. And when we consider that many of the other large mammals died out as humanity spread throughout the world, it becomes apparent that the horse found a survival mechanism by partnering up with humans, and humans felt themselves empowered. They have been an integral part of our species' history, and we of theirs. Scientifically speaking, it is not strictly a symbiotic relationship like that of a lichen—in that the algae and fungi cannot live without each other—but our intertwined histories do have a high degree of co-dependency. Consider the following quotes from equine historians:

> Horses were integral to our existence long before we had advanced culture. You might even say they gave us human civilization.[9]

> As a species, horses could not have survived without human intervention.[10]

We are so deeply interlinked that it's difficult to view human history without including horses. It's easy to overlook, but the truth is that horses are far closer to us in terms of a shared behavioral and cultural history than we may immediately think. Far more humans have cared for and loved horses than have cared for and loved our closest biological cousins, the chimpanzees. The reason for this is all tied up with our kinesis and theirs. But the significant change was not in harnessing their strength to drive wagons or ploughs as much as it was in horsemanship, riding atop.

Early cave paintings—if they can tell us anything about horses—show us that horses differ from us in their relationship with food. Many paintings show them grazing or fleeing from the hunt. The prehistoric painters didn't need much knowledge of neuroscience to know that they are hardwired as prey (rather than predators) compared with ourselves, they only needed to look at the placement of their eyes.

Horses were traditionally thought of as an animal with poor vision, but we now know that in fact it's very advanced. The placement of the eyes means that they mostly have monocular vision, but there is an overlap for about sixty-five degrees where binocular vision occurs. Their field is very wide, the only blind spot being right where the rider sits. Both us and them have a very similar understanding of motion, as evidenced by one Canadian equine psychologist, Brian Timney, who discovered that horses perceive some aspects of vision and movement in a way surprisingly similar to humans.[11]

Timney found that horses are able to experience *motion parallax*, which is how we perceive relative distance from the way objects pass by through our field of vision at different speeds. Farther objects appear to pass slower, although we cannot discern the distances themselves without binocular vision. And motion parallax depends on movement in the observer. Given that horses are relatively lacking somewhat in binocular vision, they achieve this by relying on their kinesis. They see better when in motion, which is why keeping a very tight rein isn't good for the horse's sense of space.[12] Being cursorial like ourselves, there is a dependency upon movement.

Notwithstanding that the human predatorial eyes are close together, providing an extremely good perception of depth, horses have been shown to sense perspective from a two-dimensional image, just the way we can look into the depth of a painting.[13] Perhaps horses could appreciate the depth in some art depicting them in Chauvet, da Vinci's distant blue hills, or even penetrate the flatness of a Rothko to sense its soul. Regardless of their aesthetic sensibilities, horses have a deep relationship with space and their motion within it.

Humans and horses have united the two different types of brain—that of predator and prey. Over many years, the language of horsemanship has developed between both on the understanding that horses communicate in a subtle way—via body language. They may flick an ear, or flatten both to guard their food, for example. Human and horse together represent the only known union of these neurological entities as a joined force.[14] Knowing that we could never force such a powerful animal to do anything against its will, humans have developed a method of befriending horses, so that they willingly cooperate. Highly sensitive to their environment, their direct motor responses perfectly complement the human neurological capacity for reason and strategy (found in our massive prefrontal cortex, something that horses lack).[15] Together, we have the best of both. The rider can gently squeeze with their thigh to signal a change in movement and it happens instantaneously. They can take a moment to consider options and get immediate results. Horses have the speed that humans can only dream of, and riding atop a saddle allows one to see farther with the binocular vision that the horse makes up for by keeping its deep monocular vision on the horizon. Riding a horse is a gamechanger in the story of human motion.

Yamnaya Horizon

There are some conflicting theories on how the domestication of the horse came about and going into them is beyond the scope of this book. However, some things within the quagmires of history are clear. One of the first cultures known to have managed this domestication was the Yamnaya, in the Eurasian Steppes, where Ukraine and southern Russia are today. They were a culture who lived off animal herding and raiding settlements and their reign was approximately from 4000 to 3000 BCE.

It was on these open, expansive grasslands where man and horse united, where the natural temptation towards developing speed must have caused great exhilaration. Once they managed to ride, the Yamnaya must have seen the enormous potential of this new way of moving. It's likely that they are the single historical source of horsemanship for reasons that will soon become clear. Their horse-riding prowess aided them to become so powerful and widespread that their territory covered a significantly larger area than any other culture before them.

One of the immediate impacts of horse-riding was that herding could be done on a much larger scale. It was a step towards industrialism in that the amount of already domesticated sheep that could be managed quickly doubled. Before, the herds of sheep, goats, and cattle were near their maximum at about 200 in number, but now could even be as large as 500. Larger herds meant the occupation of more space, leading to tribal conflict becoming a common occurrence and the complications that follow naturally from these types of situations. We can even trace the beginning of political alliances and trade relations back to this time—a Bronze-Age globalization, if you like—the Yamnaya expanding trade and using horses in battle situations to such effect that they were still being used in the First World War.[16]

Their second great innovation was the horse-drawn wagon. Nowadays a seemingly slow and meek type of vehicle, it was in fact a great aid for the businessmen of the time. And horseback riding in tandem with the ability to carry loads in wagons was to have deep and lasting consequences. As one of the experts in the field says, the Yamnaya "is the visible archaeological expression of a social adjustment to high mobility—the invention of the political infrastructure to manage larger herds from mobile homes based in the steppes."[17]

Historians sometimes borrow the term "horizon" from geology to denote a culture covering a large area. Just like a horizontal sediment in rock, a horizon shows how a given culture defines its time. The Yamnaya horizon is the first cultural horizon, and they gained their power through a new, cutting-edge form of motion. The form of extending one's kinesis by using equine power was the central characteristic of these centuries, just like the way that nomadism defined humanity before the Neolithic Age.

The Yamnaya got their name (meaning "pit grave culture") from the small, box-like graves and dirt mounds that covered them. They buried their horses in a similar way,[18] a practice which lasted here and there until the Benedictine Monks pushed it out sometime before the year 1000 CE.[19] Perhaps the Yamnaya understood the horse to be the intelligent animal that it is, not unlike the cave painters. Or perhaps the buried horses were to ferry the dead to the afterlife, like those found in the Viking Oseberg ship in Chapter 4. Probably, we will never know for certain, but it is a compelling hypothesis as wagons too were buried in a similar way. These have been found in Southern Russia near the Caucasus mountains. A cluster of 120 "wagon graves" have been found here, with mostly deliberately dismantled wagons (like the Khufu ship found at the Great pyramid at Giza), the four wheels demarcating the corners of the grave itself. Chariots—a later development using a spoked wheel and more advanced mechanics—have also been found in burial pits, from the Ural steppes to as far away as China, dating from just after the Yamnaya Horizon to the Chinese Spring and Autumn Period (770–481 BCE).[20]

So far, we have graves for each of the species of extended motion, boat, horse, wagon, and chariot. It's as if the extension of bodily motion goes beyond this life, extending into the otherworldly and mysterious regions of the greatest mysteries of the human experience. Given that so many cultures have left graves with vehicles, it at least shows us that movement was an important aspect of life (or life and death) to our prehistoric and early civilized ancestors. The fact that these are remarkable and fascinating finds equally demonstrates how out of touch we have become with kinesis, through the totality of sedentism in modern life.

The reason why the Yamnaya are thought to be probably the first horse riders—other cultures not learning this by themselves is pointed toward by archaeological remains. Yamnaya bones have been found dating back 5,000 years, with the effects of horse-riding discernable in the riders' bodies.[21] Comparative linguistics shows the rapid expansion of the Indo-European languages as coming from the Yamnaya expansion and covering an area from Europe to Mongolia.[22] For a single language group at the time to spread so widely, it leads some credence to the phrase "globalization" to refer to the impact their culture had. The spread of revolutionary ideas (of which more will be said in later chapters), particularly within the context of sedentism would make the change not only from prehistory to history, but from history into world history, all done from the saddle.[23] But the most conclusive evidence is in our DNA, or more specifically *some people's* DNA. Yamnaya genes have been found overwhelmingly—a figure close to 100 percent—in the Y chromosome of Bronze Age male bodies in Spain and Portugal (you may remember that the Y chromosome appears only in men).[24] This was further backed up by studies following the transmission of the Hepatitis B virus in prehistoric populations.[25] What this suggests is that the Yamnaya came as raiding bands of men on horseback.[26]

One of the consequences from these invading forces was a new culture, one which was to lead and dominate European thought until today: the Greeks.[27] In a later period of ancient Greek history, horsemanship would be dominated by those of the northern area of Thessalia, their cavalry becoming one of the determining forces of Alexander's army.[28] But before then, the Yamnaya invaders must have instilled great fear into this Bronze Age civilization. In fact, they did so to such a degree that the Greeks mythologized the horse rider as a *centaur*, a horse with the upper body of a man. Their name derives from *Kentauros* ("bull killer"), which comes from the Thessalian bull-hunting and herding lifestyle, according to the fourth-century (BCE) Greek writer, Palaephatus.[29] Mythologically, the centaurs were known to have been troublemaking thieves, often drunk and violent, characterized as uncivilized nomadic brutes, with few exceptions.[30]

Interestingly, centaurs have popped up in other scenarios too. They have appeared in cultures as far away from Greece as Russia and India, although there they were known to be less aggressive. But in the Spanish accounts of Cortés landing in Mexico with horses in 1519, the reaction from the locals was the same as how the Greeks had characterized the centaurs. The account of Bernal Díaz del Castillo, one of Cortés' soldiers, mentions how the Maya people saw that the horse and its rider "were one creature," causing them to flee in fear.[31]

A New Invention

You may remember the importance of bipedalism to the human being, how it defined us in our motion, communication, and hunting for food. Once the legs—the strongest part of the human body—were united with the horse, the centaur became concrete as an entity. And this came with the stirrup. A problem with the early leather stirrups was that a fallen rider could too easily become tangled and dragged, but a sterner material such as wood wasn't strong enough for the rider's weight when they climbed up. So, until metal stirrups came along, horse riders didn't manage to get the most out of their legs while riding, as they had to squeeze inwards with their thighs to stay seated. The metal stirrup provided far greater stability, balance, and control. Although the rider's feet are off the ground, when stirruped they are fixed to the horse via the strength of their legs, uniting themselves in a sort of six-legged warrior.

Historians are uncertain as to precisely where it was invented, but are sure it is a central Asian instrument. In yet another horse grave—this time in Mongolia—a saddle and riding boots were found belonging to a woman. Her kit was fitted out with two modern-looking brass stirrups dating to the mid–900s CE.[32] These small metal units were fastened very well to the saddle, meaning that this rider no longer had to squeeze with her legs to remain seated.

Because of the stirrup, the Mongols could ride for longer, expend less energy, and the superior stability allowed for a freedom of hands which led to their development of bow-and-arrow skills. They were known to retreat as a strategic battle play because of their ability to shoot arrows behind them. The metal stirrup allowed lances and other weapons to be better fixed to the horse's momentum, and it was far harder to unseat a stirruped rider.[33] The incorporation of the rider's legs was now part of the new kinetic phenomenon. Looking into the past kinetic history, the two free hands atop a horse is reminiscent of the running biped, enabling them to carry water, or weapons. The Mongol riders, then, represent a remodeling of the human form itself. Their agility has led some to describe them as the forerunners of fighter pilots.[34]

As horsemen and horsewomen, they had another factor for their advantage over any other culture at the time. That was the particular breed of horse that they used, called Przewalski's Horse (named after the Polish-Russian explorer that first described them in the West). These creatures were and still are the toughest horses in the world. They have a unique ability to graze when the grass is covered with ice or snow, and don't need feeding every day, like most others.

The Mongol empire extended laterally the whole way from Europe to Korea. Longitudinally, it stretched from Siberia to India, and was claimed with enormous speed. The concentration of power from this mobile advantage was so great that Genghis Khan's army of 100,000 horsemen carved out an empire the size of Africa in just twenty-five years, twice the size of any empire that preceded it. It would take more than seven hundred years until the British empire could surpass it. Its power was such that the burial place of Genghis Khan was kept a state secret by the Soviet government out of fear of a Mongolian nationalist uprising. Even the area surrounding it was sealed off by some hundred square miles.[35]

Typical of European-centered history, the "other" is represented as a brutal, barbaric, simple-minded people, and this is no different with the Mongol invasion.[36] In spite of the violence that comes with every territorial invasion, the Mongol empire was the first in history to establish diplomatic relations including diplomatic immunity, not meddle with the internal affairs of the peoples it had invaded, and give religious freedom.[37] Religious leaders, doctors, scholars, and teachers all worked tax-free.[38] Some of these social advantages are sadly lacking in many states today. Theirs was an empire with a keen focus on infrastructure, as evidenced in their remarkable postal system. Messengers were supported by a vast network of relay stations where they could rest, leave off an exhausted horse for a fresh one, and get fed, a method which lasted right up until the era of the steam train.[39] We could say that the incredible kinetic power given to them by the horse led to the movement of information, seeding today's information age. This is where what moves is intangible, but nevertheless the kinesis is there, in pure form. Many of the power structures of the world today can be related to the power of information, just like the power of horse-driven kinesis in the Mongol era. As McLuhan tells us, "any new mean[s] of moving information will alter any power structure whatever."[40]

Once the metal stirrups came into use in Europe, a new form of combat arose there, leading to a vicious cycle of ever-nastier weaponry, and increasingly resistant armor, leading in turn to further upscaling of weaponry and so on. According to one historian, the European feudal class structure arose from the stirrup coming into use in Europe. Now that horses could be used so much more effectively, the demand for them rose, giving birth to a new class of armed horsemen.[41] Unlike the relatively egalitarian horse-warrior society of the Mongols, the European knights were a male elite which dominated during a time where centralized governments did not yet exist.

Space

Under Genghis Khan, human motion had been extended systematically under the institution of a state for the first time, and it had an integrating effect. Moreover, it was the first instance of uniformity of kinesis throughout, given that the "centaur" was the fundamental unit of the empire. They were able to do all this without redefining the space that they moved through. Many cultures have "striated" space (meaning marking it by boundaries such as irrigation channels, jurisdictions, roads, geometry etc.).[42] But the space that the Mongols rode through was smooth and natural. Like a Polynesian sailor, the Mongol warrior was not bound by single-lined directions. Theirs was an open plain, which they trod with great speed, thus echoing the aphorism of Sun Tzu that speed is "the essence of war."[43] According to one influential cultural historian, the throwing of spears from horseback is more about the interruption of the opponent's movement than anything else, concluding that "the aptitude for war is the aptitude for movement."[44]

It has been argued by some that the event of mankind getting on top of a horse changed the very consciousness of humanity.[45] For certain there is the intimate neurological pairing up with our four-legged friends as discussed earlier, but that aside, people had never traveled at such high speeds, or covered such vast tracts of land in so short a time, dominating the steppes, and from so high an elevation. It's a totally different experience from walking, running, or sailing. And they began to command—through kinesis—the vastness of this world. It would be impossible for this not to affect their culture, philosophy, and even spirituality. The domestication of this animal—more than any other—was perhaps the first mastering of something so wild and powerful in nature.

And so, the horse shifted from being the other part of a symbiotic relationship with humans to a tool for territorial expansion. It provided the main transport in cities right up until the end of the nineteenth century when motorized transport stormed the world. Indeed, up to a little over a hundred years ago, the speedy form of horse and carriage transport was favored in urban design, the city avenue being specifically made for this form of motion. The long, straight roads are designed for this, rather than for commerce or pedestrians.[46] Once fossil fuels began powering modern societies, the horse receded in use, almost as quickly as it had arrived, to become a symbol of power and wealth, reminiscent of antiquity—a pastime for the upper class. Horses had been the raw power for much of the industrialization and capitalization of the world. In the classic model of industrial utilitarianism, once spent they were cast aside, or else existed simply to gratify the owner's "sense of aggression and dominance," in the words of the economist Thorstein Veblen.[47] Condemned not only to vanity but also a form of rural nostalgia, the horse is now used as entertainment, or for the flaunting of wealth.

Perhaps they will always captivate the human mind aesthetically. The horse in motion, inspiring the Chauvet painter, has done this for eons. Its beauty would continue to influence some of the most impressive feats of humanity, not just a postal service in the thirteenth century, but its natural size would determine the width of railroad tracks some six centuries later. This followed on from the Roman standard of constructing a road to be twice the width of a horse's backside to accommodate a team of two animals to carry a wagon or imperial chariot (standardized to 4ft., 8.5in.). More recently, the size of the solid rocket boosters (or SRBs) which provided additional thrust during the launch of NASA's space shuttles, was determined, albeit indirectly, by the same standard—twice the width of a horse's backside. Why? Because the company who made them, Thiokol, is located in Utah but the shuttles were to be launched from the Kennedy Space Centre in Florida. To get them there—by rail—one has to go through a tunnel, built to these same standard sizes.[48]

Muybridge

Perhaps the most interesting development of the human-horse relationship came out of an experiment to settle a wager in 1872. The disputed point was whether or not there is a moment in a horse's gallop when all four feet are off the ground. This is something that the human eye cannot see because of the speed. An English immigrant to the USA, Eadweard Muybridge had been working on cameras with shutter speeds up to a 1,000th of a second, and devised a method of photographing a horse while galloping, by a series of tripwires.[49] Each one was set to trigger an individual camera which was aimed to look along the wire itself. The result was a beautiful set of sixteen blurless photos showing an entire cycle of the horse's gait. Not only did this spark a photographic career for Muybridge,

Fig. 9. Artist's reproduction of the sixteen images of "Occident," from Eadweard Muybridge, 1872.

making hundreds of "motion studies," but it contributed hugely to the development of cinema.

By understanding motion as a series of still, stationary moments, cinema was born. The number would increase from sixteen to twenty-four photographs or "frames" per second which would be enough to trick the human eye to see them as continuous. The moving picture camera and projector is one of the most powerful uses of the wheel ever devised. Not going anywhere, it is pure motion. Just like the Mongol postal system served the motion of the intangible message at the cost of interchanging horses and messengers, cinema channels information through light over the spinning wheel. There is kinesis, but no speed of any object really going anywhere. The people and things which have been photographed have had their kinesis taken from them, leaving them suspended like the horse in the air only for the projector wheel to give it back.

It is the most powerful medium of communication known, taking control of the entire field of vision and hearing in some cases. Its purity—capturing the stillness and ironically creating motion from it—is made from nothing more than light, the speediest phenomenon in the universe. McLuhan remarks at how both the wheel and cinema arose from the observance of feet. "The wheel," he says "that began as extended feet, took a great evolutionary step into the movie theatre."[50]

Through its dreams and innovative ideas, we sit still in movie theaters and are transported without moving an inch, traveling in arrested imagination, rather than through space. Our kinesis is taken from us, and placed in the hands of a pure motion machine, leaving us moving within the realm of impressed images on the mind. Space itself becomes secondary; instead of us moving, strange and far-off lands come to us. In the sealing of the demise of the horse, the drive of industrialism and its motion led to this new medium. Fittingly, one of the dominant forms within that medium of the first industrial cinema culture is the "Western"—where the horse exists as a nostalgic memento in a narrative of colonization. It is a total and bizarre subversion. The ultimate union of horse and wheel, we might say, is not the chariot but the horse in cinema.

Countless thousands of years ago, men, women, and children likely sat around fires in caves enjoying the flickering of images. They did not do this in isolated incidents, as other different examples have been found in various caves in Europe of small stone slabs with images held very close to the flames.[51] They, like us, were captured by the movement of animals as shown in cinematic narratives. But the narrative of horsemanship itself is one of *empire*, from the Yamnaya to the Mongols and later the Spaniards in America all the way through to the intangible empire of the Western genre and other dreams that affect people all over the world.

Between the demise of the horse and the replacement of it by private motorized transport, along came a particular little invention which straddled the two. It was a very simple device, costing a mere fraction of either the maintenance of a horse or the purchase of a motor car. At precisely the same time horses were doing their last rounds on the streets in the 1890s—and a historical hair's breadth before the mass-production of automobiles—two equally sized wheels were put in a linear sequence.

Chapter 7

Two-Wheeled Liberation

> *As I turned towards the coast, and settled to the rhythm of pedaling, I experienced an exaltation I have never forgotten. The vigour of my body seemed to merge with the eager abundance all around me and in an almost sacramental way I became totally aware of myself as a part of nature.*[1]
>
> —Dervla Murphy

Often referred to as a "trusty steed," the bicycle is the most efficient and ecological form of transport known to mankind. It is far more sustainable than a horse, only needing cheap maintenance from time to time. A good quality bicycle, if maintained well, can easily last as long, often longer. And rolling on a pair of well-oiled wheels can also bring us closer to physical awareness and beauty. The correlations with other aspects of life seem to be endless, and range from the pragmatic to the spiritual.

The union between rider and bike obviously does not have the neurological wonder of horse-riding. But it is movement-orientated, and movement-dependent. Our balance is suspended by forward motion, at the same time that the uprightness of the vehicle is maintained. Both rider and their instrument are co-dependent, with motion as the fundamental condition. Without that kinetic action, we very quickly fall sideways, so we're condemned to "keep on keeping on." Maintaining stability necessitates a particular amount of concentration, and balancing can sharpen one's mind.[2] And at higher speeds, more concentration is developed, and this is only one of a plethora of health benefits.[3]

Within the world of cyclists and bikes, knowledge of mechanics is always shared rather than kept as monetizable like in other spheres. It is necessary to know some of the basics, and the elegance of the cause and effect of the mechanical world is often reflected in the reverie and passion in sharing information on bicycle mechanics. Indeed, both pneumatic tires and ball bearings come from the world of bicycles.[4] There is a camaraderie that is felt among cyclists. If one is stopped for some reason, another passing cyclist will invariably stop and offer to help resolve a technical issue, or make a gift of a spare inner tube. Through any other medium of motion this would be an uncommon act of generosity, but it is normal for cyclists to behave in this way. Cyclists tend to stick up for each other, whether in rural or urban areas, in tacit understanding that they are all cut from the same cloth, as if they form a single tribe within the urban context by virtue of the way they move.

Cycling connects with the kinetic urge, in a special and unique way, as can be gleaned from the quote at the beginning of this chapter, which comes from one travel cyclist's autobiography. Being atop the saddle allows us to feel the air as we slip through it, "as a part of nature." Despite the use of a machine, cycling still presents the elements to us, and is of course kinetically powered by our bodies. That self-sufficiency brings with it an association of freedom. One can travel the whole world on this machine; and many have done so. Not much more is needed than the bare minimum attached to the back carrier. Landscapes pass slowly enough to be deeply absorbed as if riding through a painting, and the exercise always means a guilt-free sampling of the local cuisine.

In Chapter 1, I wrote about *motion being culture*. I suggested that it's possible to read into what discourse might emerge from cultural phenomena. For example, European oil painting often shows off the possessions and prestige of those who commissioned the artist, the discourse here being that materialism and social rank was important in that culture.[5] It is no different with any other human expression, and kinesis is exactly that—human expression. But this is more a poetic interpretation, rather than an epistemic one.

If the horse was the kinetic enabler of empire, then what is the discourse of the bicycle? It is like the horse in that it allows individual travel at a far greater reach. But because it's more accessible and affordable, it is more *democratic*. The bicycle is a modern version of the horse, in the era of steel and mass production, as part of the kinesis of the modern-era city, and is billed to have a big impact in future city planning as an antidote to congestion.[6] It exists in history only after the revolutions of France and the United States, settling into the time of burgeoning republics. It is self-powered and unrivaled in its efficiency. A fit adult can travel 100 miles or more on a bicycle in a single day. Because of these characteristics, the bicycle would have a massive emancipating effect for some, and was often referred to as the "great leveler," or the "democratic machine."[7]

Prototypes and Safeties

The origins of the humble bicycle are mooted. Various inventions have been later shown to have been hoaxes, including one supposed design by no less than Leonardo da Vinci.[8] The first patent was for a velocipede (also known as the "dandy horse") by the German Karl von Drais. It incorporated a steering front wheel in its wooden frame much like a modern bicycle, but with no drive mechanism. The rider would drag themselves along with their extended feet, and toes gripping the passing ground. Described by its inventor as a "running machine," the patent dates from 1817. Velocipedes enjoyed a brief success during some years where oat harvests were poor which made horses quite costly.[9] From velocipedes to the "ordinary" bicycles of the Victorian era, all possible shapes and sizes and designs were attempted and devised by many inventors from many places.

Until the mid–1880s, the most successful of these devices was what's now known as the Penny Farthing, but known then as the ordinary. This awkward, even monstrous beast consisted of a single giant wheel reaching to about the height of a person followed by a tiny stabilizer wheel. It required great skill and athleticism not only to ride it, but even to get it going, which one would need to do before clambering up on top. The ordinary was a man's bicycle, apparently because of the clothing worn at

Fig 10. Four early bicycle models, clockwise from top-left: Velocipede, step-through, safety, ordinary.

the time. It's simply not possible to mount a wheel of six-foot diameter in a Victorian dress. The idea of a woman in anything other than a dress was unimaginable in those times, something that had yet to change, but change it did.

The constantly evolving shape came to a halt when the "safeties" came to the mass market in the late 1880s. John Kemp Starley's *Rover* safety of 1885 had a revolutionary chain drive which transferred the pedaled energy to the back wheel. This was a crucial change as the ordinary had its pedals mounted directly on the front wheel's hub. The safety allowed for the front wheel to steer the vehicle without hampering the feet. It was closer to a velocipede in size, with two equally sized wheels, the rider being able to put their feet on the ground when stopped. The ordinary's inequality of practicality was elegantly reflected in its glaringly unequal wheel sizes. The relative ease of the new machine in its handling meant that anybody could use it.

Despite intentions to allow for a rider to wear a skirt or dress, the "step-through" frame did not bring about the demise of the elaborate Victorian dresses, it was the safety. This design was rigid and strong; where the step-through, although more comfortable to mount, was considerably heavier. Its need for extra thick tubing in lieu of the absent top tube made for a lower center of gravity, making it feel heavier too. Once the safeties rolled in, clothes began to be questioned. Women's clothing at this time in the USA and UK was hugely restrictive and much or more so than the high heels of Chapter 2. Corsets were made of steel, the very same material as the bicycle frames and parts themselves, and tied as tightly as possible, giving rise to the Victorian image of the fainting woman. Their total lack of exercise served to promote the myth of women being weak, creating a vicious circle of genuine weakness leading to a further evasion of exercise and so on. One of the early opposition groups to this was the "Rational Dress Society," who existed to make things a little more reasonable. Their 1888 decree of a woman's underwear needing to be less than seven pounds in weight is illustrative of the norms against which they were campaigning.[10]

The appearance of this brilliant new vehicle led to a bike craze in the 1890s. By 1893 there were a million bike riders in the U.S.,[11] and in the year 1897 alone, one in thirty bought a bicycle.[12] They were affordable, durable, and—unlike corsets—fun and easy to use. They were adopted by all types of people ranging from seriously competitive velodrome athletes to "fair-weather" cyclists. There was sufficient demand for them to cause one immigrant living in Chicago to set up a mass production unit. Adolph Schoeninger's company, the Western Wheel Works, became known as the "Henry Ford of the bicycle," although his assembly line manufacturing plant preceded Ford's by ten years.[13]

This was a time when the reaction against the treatment of women as second-class citizens in American and British societies was being severely challenged. And so, the women's suffrage movement became intertwined with bicycling, not simply because of historical coincidence and clothing reform. These brand-new machines were instantly caught up with the politics of the time because of the spatial freedom and independence they afforded the rider.

Suffrage

As the cycling craze continued, more and more women took to using bicycles. The "new woman" was one on two wheels, this phrase becoming synonymous with the bicycle.[14] Well-known suffragettes, such as Frances E. Willard, Susan B. Anthony, Emma Eades, and Lucretia Mott relentlessly promoted cycling as a means to independent mobility. Bicycle mechanics manuals were published, specifically for women, with the intention of promoting self-reliance.[15] The oft-quoted remark by Susan B. Anthony from 1896 claims that the bicycle "has done more to emancipate women than any one thing in the world. I rejoice every time I see a woman ride by on a bike. It gives her a feeling of self-reliance and independence the moment she takes her seat; and away she goes, the picture of untrammelled womanhood."[16]

What it came down to was that a certain group of people, living within cities, made the realization that mobility was key to their rights. It is a basic right, because moving is what we are born to do and following one's kinetic urge is the expression of human fundamentals.

Women had often been subject to harassment on American public transport. They were routinely sexually assaulted by conductors or by drunk male passengers on streetcars.[17] So, it is not surprising that when modern bicycles arrived on the scene, this new, cheap form of transport-for-one became hugely popular with women in particular. It was an instrument to bring freedom and also independence from an oppressive force for those who adopted it. As one commentator says, "You can give a man a fish, but you should give a woman a bicycle."[18] Effects of this new species of kinesis in society were such to prompt the U.S. census report of 1900 to claim that "Few articles ever used by man have created so great a revolution in social conditions as the bicycle."[19]

Our kinetic urge to explore began to manifest itself in the suffragettes' push for independence of mobility. Bicycles started to leak out of the cities. Their efficiency was demonstrated many times by women eager to show how sustainable their kinetic self-reliance was. There was Tessie

Reynolds, a 16-year-old girl who cycled over 100 miles from London to Brighton and back in eight hours.[20] Elizabeth Robins-Pennell was a popular writer and artist in the 1890s whose *Over the Alps on a Bicycle* describes the new woman having "all the joy of motion ... independence and power."[21] Also, there was Fanny Bullock Workman who wrote about her experience on the roads of Spain, Algeria, and India.[22]

Going Solo

Perhaps there was no woman on a bicycle more famous during those years than Annie Kopchovsky. This middle-aged mother-of-three from Boston circumnavigated the globe to great fanfare in accordance with her natural talent for show-womanship. The legend goes that she nominated herself to settle a wager made between two men over the point of it being possible for a woman to travel around the world with a bicycle. Additional caveats were set that she would also have to earn $5,000 while doing her trip, and complete it in under fifteen months.[23]

On one of her first days, she was approached by the Londonderry Lithia Spring Water Company, and made her first $100 by hanging an advertisement for them and adopting the surname Londonderry, which is the name under which she became known. It was 1894 and this would prove useful to conceal her Jewishness at a time when anti–Semitism was steadily on the rise. Five hundred people showed up at her first day's pedaling in Boston. Her charisma was to be a great aide in gaining further advertising contracts through the whole length of her trip. Her clothing became more masculine as she traveled—gradually moving from Victorian dresses to men's suits. Interestingly, many questioned if she really were a woman.[24]

In bringing movement to the fore of expression, women had increasingly been wearing bifurcated pants, called "bloomers" after their inventor Amelia Bloomer, an item of clothing single-handedly popularized by Kopchovsky.[25] This type of clothing had been typical of lower-class women before bikes became popular, and now was something which came to represent the new woman.[26] They were the perfect, cheap alternative to the cumbersome dresses to swing the leg over the cross bar, and not have one's clothing being interfered with by the drive chain mechanism. The very absurdity of the Victorian style dress to a twenty-first-century individual was predicted by Willard, when she wrote that the "wardrobe of the rider will make the conventional style of women's dress absurd to the eye and unendurable to the understanding."[27]

There was also the inevitable crass misogyny aimed at a woman's genitals on the saddle. Some thought that the bicycle was an agent of sexual stimulation, and therefore was "morally corrupting."[28] This myth already had a history from longer-standing similar ones surrounding saddles of horses. Women had been expected to ride "side-saddle," (both legs to one side, and facing sideways) that is if they could ride at all, it being an "unladylike" activity. The largest effect of this was to deny the woman the use of the stirrup, inhibiting her efficiency, balance, and control, by both legs hanging to one side. Since as far back as the thirteenth century, European women were made to ride side-saddle. One Franciscan friar who traveled to the Mongol Empire during the reign of Genghis Khan, gave an account of how they rode there. He found that the difference in mobility and clothing between their culture and his to be remarkable.

> Girls and women ride and gallop on horses as skillfully as men. We even saw them carrying quivers and bows, and the women can ride horses for as long as the men… [The Tarter women] drive carts and repair them, they load camels, and are quick and vigorous in all their tasks. They all wear trousers, and some of them shoot just like men.[29]

But one simply cannot ride side-saddle on a bicycle, and so women took to them the regular way, and subsequently stopped riding horses side-saddle too.[30]

"Bicycle Face" and Other Dangers

As can be expected, women's new-found freedom on the saddle induced a backlash from society. Not only were they independently mobile, but they were now beginning to reject the clothes that prevented them breathing properly, all because of bicycling. Unfortunately, attacks upon women cyclists included everything from the simply patronizing to the outwardly vicious.[31]

In an incredible piece in the *New York World* from 1895, medical professionals, it was claimed, were being "kept in the dark" about the dangers of cycling. There was a conspiracy behind this which "misled [them] into an overfavourable or overconfident view of cycling." Whereas in truth, this piece continued to say, cycling is the bringer of disease to women. It causes "utter collapse," "mental excitement," and "the recurrence of acute symptoms so long as the bicycle was used." And if this wasn't enough to scare one off her bicycle, there was also "bicycle face," which was due to a swelling of the throat coupled with exhaustion from riding. Once you got it, it never went away, and of course it was only a women's disease.[32]

Although the American bicycle craze was a hugely popular phenomenon, it was nevertheless seen by economists as a threat rather than a boon to the workings of society. Since its inception, cycling has met with opposition, possibly due to its efficiency. In 1896, the *New York Journal of Commerce* estimated that cycling was taking $100 million out of the economy every year.[33] It is probably a complex mystery, but it may not be a coincidence that an invention that was emancipatory for both women and those who couldn't afford other means of transport was attacked in this way from its very beginning.

It would seem that the backlash against cyclists that happened during their first popular emergence has continued since. The individualistic, viable, and cheap independency of movement is something that could be seen as abrasive against social order. It is a political stance, as much as one about the environment, gender equality, public space, or anything else. As one opinion piece says, "the urge to push a woman off her bike while calling her a mug and shouting in her face has little, if anything, to do with road hogs and red lights."[34] Bicycles allow their riders to be free from a technocratic elite, as their maintenance is low. Backlashes against new transport means have often occurred, but normally fizzle out as we shall see in the coming chapters. Not so with cycling.

To many motorists in the cities of developed countries today, cyclists take freedoms that are not theirs to take. Often, we come across discourse that shows cyclists to be lawless, careless, and dangerous, but never mentioning the other road users who also behave like this. Even when this doesn't impact the motorists themselves, the cyclist is seen as a cheater. One cyclist can too easily paint a picture of all of them, and for some reason cyclists are collectivized, but pedestrians and drivers are not.

A British study found that although there are some who do break the rules, the stereotype of the cyclist-cheater is far from representative. In road accidents involving bicycles, police found that cyclists were at fault from 17 percent to 25 percent of the time, but drivers were so in 60 percent to 75 percent of all cases.[35] Another study reports that 83 percent of drivers admitted to speeding regularly, while 92 percent believed that they were law abiding.[36] There is clearly a problem in how things are being seen.

Newspaper columns, radio talk shows, and online comments are full of anger towards cyclists.[37] Torrents of anecdotal evidence leverage a confirmation bias, as with any prejudice. There are incidences of this hatred going further though, with reports of anonymous individuals laying thumb tacks on the routes of major competitive cycle events and of drivers deliberately attempting or threatening to knock cyclists over. There are also reports of neck-high wires across the cycleways, tied between trees![38] If it were against a group of people categorized by color or religion, much of this would be considered hate speech and hate crime.

Cyclists are an easily definable group and are clearly a scapegoat for venting anger. But there remains the question of why.

It looks like the jury is still out on this, although many social psychologists have shown that "fundamental attribution error" plays a part. This is where the identity of the cyclist is judged, rather than the conditions around that individual (that may incentivize them to run a red light, say).[39] Other studies point to different social phenomena such as altruistic punishment, where dismay is directed against those who break rules.[40]

Something we can consider again is gender. The road is a male-dominated environment as is the road rage which arises from it.[41] Some think that this is because women will tend to be slower on their bikes than men, who are also greater risk takers. In 2007, one urban planning researcher dressed either as a man or a woman on various outings. He found that drivers were far more aggressive when he dressed as a woman.[42] In terms of harassment, the same researcher says that "it's just textbook prejudiced behavior" based upon the classic social biases.

Today's commuter research tells us that the ratio of women to men cycling to work in many cities in the developed world is low. Only in Denmark and The Netherlands are there more women cycling than men, a figure of about 55 percent. In the USA it is as low as 25 percent.[43] But there is a morbid flip-side to these statistics: despite the gender imbalance of those using bikes within cities, there is an inverse relation with this to cycling fatalities. In London in 2015, although only about one-quarter of cyclists were female, three-quarters of cyclists who died on the roads were women. Gil Penalosa, director of Toronto-based urban planning consultancy 8–80 Cities, tells us that,"If there aren't at least as many women as men, then usually it's because cycling is not safe enough."[44]

This is also the case in less developed countries. A huge portion of women questioned in studies say that they don't use a bike because of male aggression. This was found to be the case in Ecuador, where the most common reason given for not cycling was a lack of respect for women.[45] In North Korea, an outright ban on women cycling was enforced in 1996, then lifted in 2012 only to be re-instated the following year. As if following on from the Victorian culture of one hundred years before, North Korea also banned women from wearing trousers in the 1990s.[46]

So even today, we have a structural misogyny embedded into the planning of our roads. And for those who do cycle, they are too often the victims of male rage on the road. I'm not arguing that it has been deliberately set up this way, but it is a demonstration that we can see certain things about our society by examining the way we move. All traffic interactions are social interactions as they happen in a social realm, therefore have a political discourse, contributing to a larger social narrative.

Modern Protests

More recently in Yemen, a young photographer has been continuing the tradition of challenging how women cyclists are perceived. Bushra al-Fusail organized a bicycle protest in her city of Sana'a in May 2015. Like in most countries, women are legally able to cycle, but socially it's unacceptable. Under the conditions of war, where there is little in the way of fuel, this contributes to a bigger problem. The first protest was an assembly of only fourteen women, ten of whom had never ridden a bicycle before in their lives. As al-Fusail says, "I wanted to show women an alternative way of traveling and at the same time challenge the mentality and beliefs of the men in our culture, show them that we have a right to ride the bike."[47] Being picked up by local and national media, the reaction—as with Annie Kopchovsky—was that they must have been men dressed as women. Beyond Yemen's borders, there were simultaneous demonstrations of solidarity in Cairo, New York, and London.[48]

It is replicated in Afghanistan too. In the years immediately after the fall of the Taliban regime in 2001, a women's national cycling team was formed. These riders braved constant verbal and physical harassment in a sustained effort to bring about cultural change in their country, refusing to allow the suppression of women to continue.[49] They saw themselves as ordinary young women, rather than athletes, confident that this is something that shall pass and women's cycling will, at some point in the future, be normalized.

If these trends continue, women cycling will become a normal, everyday occurrence. Better infrastructure the world over leads to more cyclists and better cyclist behavior, which in turn leads to less aggression, inviting yet more people to cycle.[50] It seems from the stories of Yemen and Afghanistan that the removal of mobility from a person even outdoes their disenfranchisement, similar to the sentiment that war is about interrupting the enemy's movement (from Chapter 6).[51] To vote is central to democracy, whereas to move is central to humanity.

Love

What is striking about Dervla Murphy's statement at the start of this chapter is that it also connects with the static urge. There is the relating with the sacred, the union with nature, and the autotelic (not goal-oriented) nature of some bicycle trips. From another perspective, cycling is about finding peace within movement, settling and accepting a particular way of life which is satisfying by the means rather than simply the ends.

Despite all the aggression, cyclists regardless of gender love being in the saddle. There is a great freedom brought about by cycling. It is often seen as a way of life more than simply a method of getting from A to B. Traveling by bicycle is so efficient as to be a thing of simple elegance. They take up just a little more space than their riders. So, the ownership of space is close to equal, unlike other modes of transport. This will undoubtably be of importance for urban policy in the future as congestion increases as a problem in our cities.[52] They are quieter than motorized vehicles, extremely easy to use—literally unforgettably easy—and beneficial for health.

Many believe bikes to be unsafe, but the causes of danger are in the surrounding environment. A bicycle is not inherently unsafe. And the exposure of the cyclist is precisely what protects others around them. A cyclist cannot endanger another person without endangering themselves. It is this fact that differentiates bicycles from other vehicles in a city. A bike is an extension of the human being, whereas a car is an extension of many more things. A bicycle is therefore much closer to a pedestrian within the urban context. You may remember from Chapter 5 that I describe the city as a catalyst for inequality when compared with the more egalitarian hunter-gatherer societies that went before them. In this context, the city has been described by some as oppressive towards women.[53] The bicycle is the kinetic instrument which challenges this system.

But the kinetic urge was not sated with this, the most efficient means of motion. Cycling federations in North America and the United Kingdom as well as other countries began a campaign for better roads at the turn of the century. We tend to think of that infrastructure as being made for automobiles, but it was pedestrians and cyclists who first campaigned for smooth roads.[54] And there was another mode of transport already in full swing in the industrialized countries, which by the time of the bike craze was rolling itself out worldwide.

Chapter 8

Industry

As for me, I am tormented with an everlasting itch for things remote. I love to sail forbidden seas, and land on barbarous coasts.[1]

—Herman Melville

So far in our story we've dealt with purely human forms of kinesis such as walking and running. We've also dealt with the extensions of motion, into boats, horse-riding, and the bicycle. With the advent of industrialism, new types of kinetic extensions became possible which moved people in far larger numbers.

Industrial types of motion are where many people travel together, with speed as a primary characteristic, and it does not rely directly on organic power. All of us are bundled into some sort of container which is of a different scale from what we discussed before. Ancient practices of traveling by raft or boat—discussed in Chapter 4—are the first step in this direction. Rafts were used as far back as *Homo erectus*, to float out on the sea. This brought them to faraway lands. *Homo sapiens* later colonized continents as far away as Australia, an incredible 30,000 years before they got to Europe.

Discovery

Floating upon the dark waters and using stars for navigation, sailors from many eras must have been overwhelmed by the night sky. When ocean-going ships were built, the demographics of the world were changed. Using newer methods of knowing their whereabouts, European explorers and conquerors began to regularly cross the Atlantic. The quote above, from the author of *Moby Dick*, suggests being drawn towards going to sea precisely because of its dangers, showing that curiosity has some part to play in all of this. Ultimately, it was through the development of industrial forms of kinesis that humans would travel up to one of the celestial objects in 1969.

Reaching Melville's "barbarous coasts" inflated our bravery, and exploded into a new era of history in the fifteenth century after Vasco da Gama, a Portuguese sailor, rounded the Cape of Good Hope on his way to India. Accompanied by a series of improvements to the astrolabe, octant, magnetic compass, and the chronometer, great leaps were made in navigation. There was also the wider availability of maps thanks to Gutenberg's printing press. The ocean-going ship, a massive, powerful machine, could take many times more cargo or passengers than a simple raft or dugout canoe. All these industrial changes together brought about the European Age of Discovery.

The important difference for our story was that for the first time there was a division between those at the helm and the passengers. In earlier sea travel, this difference was far less, or absent altogether. With ancient rafts and sailing boats there was a lesser hierarchy in the social system that evolved aboard their vessels. But in the Age of Discovery, that power belonged to a smaller number of people, the rest becoming reduced to mere operatives or cargo.

Now that the Earth had long since been covered by humans and their great variety of cultures, a new layer of colonization wrapped itself around the globe. This time it came from those who had access to different kinetic methods. The concentration of power in industrial movement is reflected in the technological difference between those with and those without it. As a grim example, recall the army of Hernán Cortés from Chapter 4. It needed merely 508 soldiers to take the Aztec capital of Tenochtitlan (with a population of some 100,000), such was the advantage of having gunpowder and a cavalry.[2] Europeans re-covered the Earth, and through a mixture of trade, genocide, and repopulation, they redefined the geographical makeup of humanity throughout the world.

And through this industrialization, the cruelty that came with it was also unprecedented. The tentacles of European power robbed the world blind, and the descendant wealth is responsible for much of the poverty gap that exists now in the twenty-first century.[3] This new industrial motion created the largest known slave trade, lasting for hundreds of years, and displacing millions of Africans. The ramifications of colonialism still make up the largest historical force shaping the world today.

Modernity

The symbol of industrial capitalism, however, is the train not the ship. The ship symbolizes the discovery which led to industrialism, through the institutions of colonialism and slavery. These institutions led to developing the necessary wealth for the steam locomotive to arise. It was here, in the developed world where the fullest expression of industrial kinesis first established itself.

As with ocean-going ships, passengers have no control over the vehicle. Train passengers do not even see in the direction in which they travel, their visual field being a world apart from that of the driver. They have quite a different vista from those in control, and that division of control of the vehicle is absolute, resulting in a kind of enforced trust of those in charge upon those in transit. This caused a common fear in the early days of rail travel and it was only made worse by the drivers' compartment being completely separated and inaccessible.[4]

The arrival of trains had a massive impact on almost every aspect of life in the nineteenth century, and when they came, it was with a bang. Railroad building erupted worldwide, with about 600,000 miles of functioning railway built between 1830 and 1900.[5] Initially, it was envisaged for the transportation of goods rather than passengers, but the new machines too easily captured the imagination of the general population, and passenger travel really took off. The countryside was now far more accessible to people, most particularly in Britain. A new fascination for the British homeland developed, and gene pools even expanded as travel became easy and affordable.

Trains brought people and places together, sometimes connecting them for the first time. With steep increases in efficiency, the machines became faster and faster. While the frantic building of rail lines connected cities with towns all over the world, other new forms of motion were coming into being, too. The ordinary was gaining in popularity, as the safety was yet to be invented. The first experiments were being carried out in combustion engines for "self-driving" carriages, which would quickly be developed into the modern automobile. Electric batteries were developed as were overhead cables for "horseless" carriages. The rate of change concerning new species of motion exploded, and the train was the powerful symbol of these times and changes throughout the nineteenth century.

The earliest trains were already far faster than stagecoaches, so the difference in traversing space and time was instantly visible.[6] This gave rise to the coining of the phrase "shrinking space and time," with the railroad knowing "only points of departure and destination" in terms of space.[7] The opening up of the countryside unfortunately came along with the carving up of it. The railroads delineated, marked out, and cut through the countryside in a fashion that was hardly visible from the train carriages themselves.

As the various towns and villages were effectively brought closer together, trains brought about an important change in the way time relates to kinesis. From the nomadic perspective, time is merely a measure of motion. This was somewhat reversed in the first settled societies, which depended hugely on time, calendars, and weather. But trains reinstated this ancient view of time, changing it from *primacy of time* to *primacy of motion*.[8] This means that kinesis was once again seen as a fundamental phenomenon of how the world works, and time was secondary to that. It was to be the beginning of a huge increase in movement being part of modern life, which continues today. We move more than ever and faster than ever, even though most of us live a settled existence.[9] And the first instance of this change was how time-keeping was affected by trains.

Clocks

The need arose for a common time in order to schedule trains between cities. During the 1880s in Britain, as trains became increasingly widespread, there was an ever-growing demand for the country as a whole to operate on a standardized time. This was to make sure that there would be only one train on any one track at any one time. Before this, there was no reason for a given town to have the same time as another, and for a while each rail company had its own time, with the defining clocks traveling on board the trains themselves![10] All of this happened independently of the Greenwich observatory despite it having been in operation for some two hundred years already. As the communication between places increased, such an atomized system was hardly sustainable, and furthermore the scheduling of trains was necessary for safety. By 1885 the world had quickly been divided into time zones.[11]

It was during this epoch that Henri Bergson—one of the best-known philosophers of his generation and later a Nobel laureate—published his doctoral thesis in 1889, becoming one of a very small number of philosophers after Heraclitus to deal with motion and time. Here, he distinguishes between what we consider to be objective time and what we experience:

> When I follow with my eyes on the dial of the clock the movement of the hand which corresponds to the oscillations of the pendulum, I do not measure duration, as seems to be thought; I merely count simultaneities, which is very different. Outside of me, in space, there is never more than a single position of the hand and the pendulum, for nothing is left of the past positions. Within myself a process of organization or interpenetration of conscious states is going on, which constitutes true duration. It is because I *endure* in this way that I picture to myself what I call the past oscillations of the pendulum at the same time as I perceive the present oscillation.[12]

Bergson reminds us that the clock *measures* time, it doesn't represent it. We, on the other hand, experience it in an internal way, subjectively. Each moment is unique not because it will only happen once, but because we carry into it the memories of previous moments. Hence, he "endures" through the previous oscillations while perceiving the present oscillations of the clock's pendulum. Bergson used the phrase *la durée*—"duration" to describe this inner world of how we perceive time, and *les temps*—"time" for the outer, objective time. We experience the passing of time in a qualitative, personal way, all under the primacy of motion, for time can only be a measure of movement.

What's remarkable about Bergson's writing throughout his career is that this understanding of time always relates very directly to the nature of our consciousness as human beings. Because of its nature as he describes it, the only way to examine it is through our own intuition. Our own interior experience of ceaseless motion and duration are the only realities we have upon which to base how we see the world around us, leading him to state that "reality is mobility."[13]

Not merely was time being shrunk by trains, but space too. It was commonly understood during the nineteenth century that railways were not only shrinking time, but "*annihilating* space by time."[14] Remarks on the subject included descriptions of whole towns fitting into streets, and both time and space being "destroyed."[15] Trains were swallowing up more and more space, as they expanded into the remoter regions of each country. These places would become homogenized in a manner similar to today's globalization. The placement of a town, for example, became less important due to the new ease at which space could now be traversed.

Trains had a mechanizing effect on the societies that used them, this being the first complex machine that was common to everyday life.[16] A culture of bureaucracy grew alongside them, with those in charge being ordered into a quasi-military, with army-style uniforms, various levels of accountability, emphasis on punctuality and so on.[17] The most important military element missing was drill training, keeping in step and time with their bodies. This was outsourced to the machines themselves, leaving the officers to operate a banal bureaucracy. Perhaps this reflects how unadaptable a system is once rails are put down, unlike the system of roads and cars which came after it. It was being able to break from the technocratic and inflexible nature of trains that probably led to their being eclipsed by the car.

Images

Many readers will be familiar with the act of staring out the window of a moving train. There is an unusual intensity to it, often hypnotic, and it can sometimes be difficult to look elsewhere. The passing landscape presents itself to us as if flat, like on a screen, suspending our attention like deer arrested by headlamps, and we behave indifferently to the sequence of impressions, disconnected from it by the glass. What arrests our attention is the consistent rhythm of movement, rather than an interest in the content of what we see. That conformity of place brought about by the presence of rail travel reflects itself in our own passive acceptance of the moving landscape as we sit and stare.

Through the window of the moving train, its first passengers saw the kaleidoscope of constant change at speeds never commonly experienced before, while they sat still. With this machine, the effect of cinema entered the public consciousness. The rolling montage had the same impact both on the train and the movie theater, almost hypnotizing its audience. Once trains had established themselves as a standard way of travel, the early innovations of cinema began. Cinema is the intangible counterpart to the train, a purity of motion made up only of light and the suspension of disbelief.

The great difference between the two, of course, is that where the montage of the train window is genuine kinesis, the images in a roll of film are static. Bergson reminds us that it is the *apparatus* of cinema which brings motion to us. As if the subjects have been frozen by the photographer, in the replay "each actor of the scene recovers his mobility."[18]

In a beautiful coincidence of history, it was a train that appeared in one of the earliest known public screenings of a movie. In January 1896 in Paris, two brothers, Auguste and Louis Lumière, who led the innovation of cinema, gave a screening of their fifty-second motion picture, *L'Arrivée d'un Train en Gare de La Ciotat*. It was to make film history as the images depicted a moving train coming directly toward the audience. The legend goes that those in the movie theater panicked in the belief that the train would hit them.[19] Such was the power of both of these new innovations. But before thinking about the innocence of that audience, just reflect on how many times a film has made you laugh or cry. Some years afterwards, the movie camera itself would shortly find itself copying the train in the form of the "dolly," where a shot is made by a camera rolling smoothly on tracks.

Here we also find a strange relationship with the horse. The first public transport—streetcars, trams, stagecoaches, and horsecars—were horse-drawn. They were being discarded at exactly the time of Muybridge's famous photographs. His series of still, unmoving shots, relate not merely to the natural gait of the animal. They relate to the experience of a passenger on a train. They are among the earliest cinematic phenomena, each frame taken from a particular position equidistant alongside the trajectory of the horse, just like a dolly camera. All other species of motion became industrialized, except the horse. It was instead made obsolete; its kinesis taken from it and being printed frozen onto the flickering film. This marks the beginning of the strange eclipse of reality by its repetition, a pattern which abounds in the world of film and of modernity in general. Landscapes become panoramas, and real physical depth in vision—also shared by the horse—becomes flattened onto the screen of the train window. Cinema—a perfection of the repetition of reality—was the extension of these photographs, perhaps in a way an extension of both the horse and the train themselves.

Space and Society

As always, rapid change brings skepticism, and many people in the nineteenth century feared the loss of their well-established way of life. People of varying cultures and classes rode the trains together, many of whom not having mixed before. To some it was socially progressive, but others felt it as a threat to their values, causing a strong backlash:

> Railroads, if they succeed, will give an unnatural impetus to society, destroy all the relations that exist between man and man, overthrow all mercantile regulations, and create, at the peril of life, all sorts of confusion and distress.[20]

Either way, rail was undoubtedly something that did bring people together in a new paradigm, doing so in societies even as large and diverse as India. Despite the later presence of carriages defined by class, it certainly did have some socially unifying effect. Other contemporary accounts describe the "living mosaic of all the fortunes, positions, characters, manners, customs, and modes of dress that each and every nation has to offer," arguing that the experience of simply being on a train shall "do more for the sentiments of equality than the most exalted sermons of the tribunes of democracy."[21]

But if it was doing turns for democracy, it was also enabling a structure of control at the same time. One of the defining characteristics of industrial motion, which I mentioned earlier, was hierarchy. Larger groups of people simply need a more complex system of organization. And just as the first settled societies needed new systems of social organization (see Chapter 5), doing so through spirituality, their modern counterparts organized themselves in hierarchical industrial kinesis. So many new methods of moving appeared over the span of just three generations: the bicycle; the automobile; the train; the airplane.

The industrial machines of kinesis differ in that the passengers have no control. Sailors were in charge of their vessel, not the passengers. Slaves were treated as cargo—dehumanized as sadly they would remain for the entirety of their lives. This is probably the earliest and most brutal kinetic instance in which people were divided into those with the control of the vehicle and those without. Any earlier form—such as a horse and carriage—had a reversed hierarchy where the passengers enjoyed the servitude of the driver. The slave ships, however, directly used this new-found efficiency of motion as a socially controlling device.

Some passengers felt atomized by the train, as if they were cargo, just like parcels.[22] People's mobility was increasingly seen as being "monopolized" by railroad companies, and industrial kinesis changed how we relate to space.[23] When traveling by train, passengers have little or no relationship with the space that they move through. Separated and insulated, they arrive at the new station disconnected from the experience of getting there. We lack the geographical relationship with space, the next best thing being the secondary reality of looking through the train window. It's as if stasis is incorporated into kinesis, we move but do so within the still bubble of the railroad car, ship's cabin, or automobile. A union forms from the desires to explore and to be still; hence the later idea of simply going for a drive and not leaving the car, just passing through the countryside, separated from it by the glass.

We have far less knowledge of space now, less feeling and intuition. It is of no consequence if we don't know in what direction we've been traveling, be it north or south. The French anthropologist Marc Augé—in his book about the Paris metro—tells us that "the ways of the metro, like those of the Lord, are impenetrable: they are traveled endlessly, but all this agitation acquires meaning only at the end, in the provisionally disillusioned wisdom of a backward glance."[24] In this formulation, movement is about the destination, not the journey, and there isn't a better example of this than with flight.

Flight

Since time immemorial, humans have looked up to the sky, marveled at birds, and dreamt of flying. For us Earth-bound beings, it is the kinetic expression of pure freedom, totally escaping our realm, and flying without any geometrical or linear restriction up into this broad, open space, free from our dependence on the ground.

The effect of the railway forced a fundamental rethinking of space.[25] From the "smooth" space that the Polynesian sailors traversed and the early human nomads hunted, this new space—as well as being shrunk—became "striated" (meaning being marked with long lines).[26] Rail measured space very strictly and the striation of space was quite literally iron-clad. Throughout history we find that human activity has this pattern of a changing relationship with space—taking smooth space and striating it through our kinesis, measuring it, setting boundaries, lines, routes, and territories. Following rail, one of the last smooth spaces left was the air as a medium of movement.

It is for this reason that the conquest of the air has so often been associated with freedom, or a higher spirituality. Winged angels appear in Christianity, and existence in the afterlife can be in bird form in other cultures.[27] Through death, we finally escape the human kinetic condition of being grounded. The great structural anthropologist Claude Lévi-Strauss labors the point that birds are generalized about far less than other creatures, their individuality is never compromised because they're seen to possess a genuine freedom.[28] There is the phenomenon of naming birds with more human-like names than any other type of creature, as if they exist in a parallel world to ours, precisely because of our inability to follow them into the sky.[29] In desperation we project our values onto them in an attempt to come a little closer to their kinesis. With their freedom of movement comes their identity, or even a supernatural origin.[30]

In the famous Greek legend, Daedalus and Icarus managed to fly, and as the often-told story goes, Icarus flew too close to the Sun which then melted the beeswax which had held the feathered wings in place. Later, the poet Ovid warns us that by creating these wings to leave the island of Crete, Daedalus "put his mind to techniques unexplored before and altered the laws of nature."[31] To alter the natural order has been one of the greatest of human ambitions despite this warning. And it is achieved through the striation of space, and through the attainment of powered flight.

Surely there is no other form of kinesis which demonstrates so elegantly the technological achievement of mankind than flight. The grace of the airplane, glider, zeppelin, or even hot-air balloon is invariably a capturing sight. More than one hundred years after the Wright brothers finally cracked how to achieve it, we still gaze astonished at these wonderful machines.

As an extension of our kinesis, the airplane is perhaps the most bizarre of all. The fluid mechanics of how a plane maneuvers in the air, resemble how we dive into water. This is where speed plays its part. Air is a thick viscous substance, but we easily forget this, mainly because on land speeds are not sufficient for us to notice. A plane moves fast enough to render the air as equivalent to our experience in water. Flight is the final use of the wheel, totally exasperating it, making it roll against nothing, once takeoff has been achieved. The ground is made obsolete, now that the medium through which the plane moves envelops it, just like a person swimming. In a way, the air becomes a version of the sea.

And so, it distantly relates to humans swimming, but is far more an extension of the bicycle. Both flight and cycling depend completely on a minimum forward speed and aerodynamic balance. Not moving forward, the cyclist won't last long before losing their equilibrium. Not moving forward won't achieve much for the pilot or their passengers. It's forward motion that generates lift from the wings of the airplane, the existence of both depending on speed. As McLuhan says, "It was no accident that the Wright brothers were bicycle mechanics, or that early airplanes seemed in some way like bicycles."[32]

This form of kinesis reached its zenith with the moon landings of July 1969 when two Americans placed the flag of their country there, claiming to "come in peace for all mankind." Humans had claimed ownership of space but it was the action of getting there and back that was the triumph—inspired by wonder, like many sea voyages.[33] These space exploration projects, such as the NASA's Gemini and Apollo programs, grew directly from flight, being manned by military pilots flying at ever higher altitudes. And here we enter into the non-tangible forms and benefits of exploration. What's gained from space missions is technological, rather than territorial like in oceanic colonialism. At the same time, the space that is conquered is intangible. Space itself doesn't have resources for us to take home, so its exploration is closer to motion for motion's sake, a much purer expression of our innate curiosity, bringing us full circle back to the earliest colonizers. The twelve people who have walked on the moon might have done so in the same spirit as the first *Homo sapiens* who left the Great Rift Valley about 90,000 years ago.

The airplane is a miraculous link between our modern cities, shrinking time and space so much more than the train ever did. We are left looking down just like at a map. It is a huge player in the formation of our globalized world today, with a single language for air communication and a single time zone for aircraft to operate within.[34] Bit by bit, our cities are becoming more homogenous and increasingly like airports. Where street traffic changes from culture to culture, airports differ much less.[35] According to Augé, places like the airport create a non-organic atmosphere, where one feels isolated in a "contractual" way of interacting with those around them.[36] The type of space of the airport oozes out into other realms of modernity such as hotels, casinos, waiting rooms, and meeting rooms in a sort of spatial globalization.

In conclusion, industrial motion is another turn in the story of our kinesis. It represents a huge leap in terms of the extension of motion. With industrial motion, one can travel around the world far more easily than when compared to the simpler extensions of motion such as the bicycle, horse, or automobile. Industrial motion operates within a different category of efficiency and speed.

While large-scale motion opened kinetic possibilities to many people, and extended their reach, flight did not do this. An estimated 11 percent of the world population have used aircraft, with frequent flyers at about 1 percent.[37] The freedom that the plane provides is one which is embedded in many constraints, including poverty and mobility.

But there's another part of this world of modernity, expressive of its hyper-individualism, something that doesn't find expression in industrial kinesis. It follows the desire to liberate oneself from striated space, to travel alone, far, and to travel freely. Early models were attempted using the steam engine, but at such a small scale it imposed great limitations. It wasn't until the internal combustion engine was invented and the cheap fuel to power it came along that the automobile changed the face of the world, our cities, the way we relate to each other, and how many define freedom.

CHAPTER 9

Autopia

Cars have become the real population of our cities, with a resulting loss of human scale.[1]

—Marshall McLuhan

Sometime in the sixteenth century, Michel de Montaigne wrote that "We are never at home, we are always beyond."[2] He alludes to our extending of ourselves to a territory outside the body, driven by our desires, and attachments. A century later another thinker, heavily influenced by Montaigne, Blaise Pascal took this a step closer to kinetic restlessness by describing the cause of all human unhappiness to be because we "cannot sit still in a room."[3] He identified the kinetic urge to move within the context of stasis, how it insidiously pervades our existence, and we cannot sit still. In the twentieth century, this was manifested by us sitting still in our "rooms"—or automobiles—to move about, our bodies themselves not moving whatsoever. The car differs from the train in that we can rarely engage with strangers, leaving us isolated from society while driving.

You may remember from Chapter 5 that I discussed sedentism in terms of living in a city. The urban world is a realization of our urge for stasis, but it's mixed into the reality of kinesis, a world where all things are in motion. The car is an expression of this too, but of an extreme form of "thing entanglement," where we are tied up with the consequences of ownership of items on a level where most advanced technology intersects with us on an individualistic basis.

Unlike the train, the car can bring us from home to anywhere we need to go. It gives more freedom of movement than the horse or bicycle, and we also regard it in a sort of personal way. Cars have become so successful that McLuhan has called them the "real population of our cities," in the quote above, and the scale he talks about is that designed not for the drivers of the cars, but the vehicles themselves. Despite the immense possibilities the car gives us, it is also sedentism at its most extreme. We can travel so freely while hardly moving our bodies an inch.

The entanglement that has been created by those who built the automobile infrastructure and which the rest of us have fallen into, is a single system—by which I mean the highways, car parks, gas stations, the insurance industry, advertising, the myth of freedom etc. Over a single century it has grown to a point of *totality*, where the presence of the car saturates its environment. It is everywhere and as McLuhan says, our cities accommodate it more than us. It is difficult to live outside of the automobile system, particularly in the United States, where there are approximately nine cars for every ten people.[4]

We might go as far as to say that the automobile is the defining object of the last 100 years, and I hope to explain why in this chapter. Seen to be the epitome of freedom of mobility, these affordable vehicles became one of the most produced objects known after the Second World War. Post-war global politics have been dominated by a dependence on oil, made so in part by the automobile. In a car we can go anywhere we please, with few exceptions and with few problems.

Some have remarked that there was no demand for these vehicles at the time of their development toward the close of the 1800s.[5] While this may be true, and the demand for them was certainly manufactured to a high degree, it's only natural that people want to see new places, travel, and explore.

Towards a "People's Car"

At the turn of the century in the United States, there were many simultaneous inventors of early motor cars, mainly using electric and steam power. In those years, gasoline provided power for less than one in four car designs.[6] From Gaston Planté's automobile in 1859 to Thomas Edison's in 1895, a large swath of experimentation took place, not unlike the early years of bicycle design. When Gustave Trouvé attached an electric battery to a Starley tricycle in 1881, it became the first model to carry people. One outstanding fact demonstrates how steep the curve of progress was: by 1899, the 100 km/hour barrier (62.5 mph) had already been broken by the Belgian driver Camille Jenatzy in the bullet-shaped "Jamais Contente" (the "never contented"), a name stunningly reminiscent of Montaigne and Pascal.[7]

The electric models were more viable than steam, the latter taking quite some time to get going, essentially being scaled-down versions of train engines. Compared to horse-drawn carriages, the electric car was quiet, clean, and odorless. Compared to the gasoline-powered models, they did not require the hand crank to start the engine, or have the complications of gears, but were limited in range. But once cheaper fuel pushed the market in the direction of gas-powered cars, a new trend was set, and numbers exploded throughout the century. By 2003, the number of cars in the U.S. surpassed that of driving licenses in the country.[8]

These days a special and deliberate effort must be made by most people wishing to be in a place where cars cannot be seen, their global population now having surpassed one billion. This alarming number is growing faster than the human population and is predicted to hit two billion by 2030.[9]

For this population explosion of cars to occur, they needed to become an ordinary and necessary item. Far from the luxury of the early automobile, a "people's car" was needed. This was the thinking of perhaps the twentieth-century's greatest industrialist, Henry Ford, the man who would be incorrectly credited as the inventor of the automobile. Many other myths have come to surround the man who took up half the world's market share of the industry at one point.[10] He is credited for being the inventor of the assembly line, something which had actually been established in bicycle manufacturing a decade earlier. Ford earned a reputation for being a kind boss after the announcement of a five-dollar workday, which, in the first half of the twentieth century was hugely generous pay for unskilled work in America, normally valued at between one and two dollars per day. He later remarked that this was the greatest money saving incentive that he had come up with due to the reliable labor that came with it, and raising working-class income thus creating more demand for his product.[11]

His people's car came in 1908 when the first Model T or "tin lizzy" was put on the market for $850, a price which would gradually get lower over the next fifteen years. The world had entered a new century, one with a wonderful technological future. Developed countries were entering a phase of late-industrialism and the suffragette movement was in full swing. Car production was taking off worldwide, with the manufacturing leader being Armand Peugeot in France, who was already producing about 10,000 cars a year when the Model T came on the market.

National Rail Lines

In its first decades, the car was a luxury item, and the majority who couldn't afford it used public transport—which was mostly electric-powered by a centralized source. For it to be commercially viable it needed three things: affordability, demand, and infrastructure. The problem is that the latter two are conditions which lie outside the realm of the automobile manufacturers. Affordability came easily once assembly lines and cheap gas were both in place, the latter in turn eliminating the electric and steam-powered designs.

Those in control of the industry had to resort to ingenious methods for demand and infrastructure to increase. From behind the scenes, a plethora of strings was pulled to direct city planning to accommodate the new vehicles, all done incognito. Between 1938 and 1950, a single company took monopoly ownership of the public transport systems of a plethora of cities in the USA.[12] Called National City Lines, Inc., it was owned by a partnership of General Motors, Standard Oil, Firestone, Mack Trucks and some others. They gradually replaced the electric trams and trolleybuses with diesel-powered buses under the false pretext that buses are more profitable. Many cities in the United States had their rail systems bought over—in part aided by a new law which was put in place in 1935 to combat the monopolizing of the electricity market.[13] The list of cities includes: Tulsa, Jackson, Kalamazoo, Saginaw, Montgomery, Mobile, Cedar Rapids, El Paso, Beaumont, Port Arthur, Springfield, Portsmouth, Canton, Butte, Fresno, Oakland, Stockton, San Jose, Los Angeles, Glendale, Pasadena, Long Beach, Sacramento, Terre Haute, St. Louis, Lincoln, Salt Lake City, Portland, Tampa, Baltimore, Spokane, and New York City. Over 100 rail systems in forty-five cities were affected.[14]

All of these cities had their municipal transport systems bought by this partnership, in the form of National City Lines, their tracks ripped up from the street, overhead power cables taken down, and even the carriages burnt on occasion. It was a systematic dismantling of the public lines, ostensibly for the replacement of trams by buses. But then, once the buses were in place, it became apparent that they were less efficient than their predecessors.[15] The buses were overcrowded, smelling of diesel, and their service was decreasing in quality, having fallen into a spiral of inefficiency and lessening demand. The streets were now ready for the automobile to take over. It was a great generator of taxes, which meant more money which could be spent on roads. This in turn increased the number of road users, generating even more taxes. Between these two circles—one vicious, one virtuous—the automobile would triumph. By the mid-1940s most of those involved in the scheme to replace the trams and trolleys had sold off their shares in National City Lines, and it shortly went out of business, leaving only some of its subsidiaries.

This oligopoly—General Motors, Firestone, Standard Oil, Phillips Petroleum, and Mack Manufacturing—were all convicted for criminal conspiracy to violate antitrust laws by a federal grand jury in 1948, for "a conspiracy limited to the monopolization of sales to a single corporate system, the City Lines defendants and their operating companies."[16] The conviction was appealed two years later, but the court's decision was upheld.

But despite this, National City Lines, Pacific City Lines (a subsidiary), Firestone, GM, Phillips, Mack, Standard Oil, Federal Engineering, were all charged and fined a mere $5,000 each—about the price of five Chevrolets.[17] The individuals involved—including Roy Fitzgerald, the head of NCL—were fined a token $1 each, plus court costs of $4,220.78.[18]

A further case was heard at the U.S. supreme court in 1957, convicting GM, DuPont, and United States Rubber of monopolizing their market.[19] But by this stage, most of the damage had been done and some even continued, the case in point being Los Angeles, where at the end of the nineteenth century, 90 percent of streetcars had been electrically powered.[20] By 1960, the final streetcar and local railway lines were being ripped up, the last suburban railway—going to Long Beach—was closed in April 1961, and the city's last streetcar lines were axed in 1963, resulting in LA now being the most car-dominated city on the planet.[21]

Fordlândia, War, and Technocracy

As the twentieth century rolled on, cars became increasingly present in all walks of life. They even made their impact in the Amazon rainforest, in an area with no roads, with Ford's bizarre jungle city project founded in 1928.[22] This was an attempt at marking his own corner of the rubber market to produce Ford's own tires for his factories in the U.S. He purchased land in Brazil, specifically for the purpose, and gave it the not-so-subtle name of Fordlândia. It was here that Ford showed a different attitude to his workers. In this town, there was a total segregation between the Brazilian and American employees, allowing only the latter access to running water. It was the closest thing to a technocratic dystopia, with all aspects of life controlled, even diet.[23] After the project was abandoned due to a blight affecting the rubber trees, violence erupted in Fordlândia forcing the Brazilian military to intervene, and the American workers returned home. In 1945, it was eventually sold back to the Brazilian state by Henry Ford II, the grandson of the industrialist, having never successfully supplied rubber to a single factory.[24]

Car manufacturing deeply affected the war effort too, particularly among the "big three" companies of GM, Ford, and Chrysler. For some it was personal, as many of the automotive industrialists were anti-Semites, including Morris, Porsche, and Alfred Sloan (of General Motors), who allowed all the Jews in his factories to be expelled.[25] Opel, a subsidiary of General Motors, allowed their factories in Europe to be "Nazified," leading Albert Speer, the famous architect and Nazi minister for production, to declare that Germany would have been unable to invade Poland and Russia without the help of GM.[26] The war showed the impact of the automobile approaching its saturation point as Ford was supplying vehicles to *both* the Allied and Axis powers during the war—the British using the Model T chassis for the military, despite the objections from Ford himself.[27]

Ford was well-known for disseminating anti-Semitic literature through his dealerships, insisting that they pass it on to their customers.[28] He received the Nazi Grand Service Cross in 1938, the highest honor that a foreigner could be awarded, and only the fourth person ever to receive it.[29] The following year Hitler received a birthday gift of $50,000 from him, and Hitler cited Ford as an inspiration in *Mein Kampf*.[30]

Continuing on from his disciplinarian attitude to the workplace in Brazil, Ford employed a fascist model of rule in his factories in the United States. Security personnel were hired as bouncers and spies, some even using tigers to intimidate the workers.[31] Not unlike the communist and Nazi dictatorships themselves, Ford's plants ran on fear in the 1930s under the security management of one Harry Bennett, formerly a boxer and sailor, whose system has been described as the "largest secret police force outside Nazi Germany and the Soviet Union."[32]

Ideas of total control of motion were adopted in the Ford factories, as a second technique of control alongside the strict rule of Bennett.[33] The blossoming industry of "scientific management"—or "Taylorism" after its inventor Frederick Taylor—incorporated stopwatches and ever-more precise work instructions. It mechanized the human body and disallowed agency in the employees.[34] Other examples of this borrowed from Eadweard Muybridge's motion studies, making very detailed maps and plans—using long exposures with lights on certain parts of the body—of how a worker ought to carry out their tasks. In the words of some of these experts, "Through the data derived from these, we standardize motion paths, motion habits, and all other motion variables. These enable us to test and classify, select and place, both work and workers...."[35]

During the first half of the last century, the car embodied many mythical qualities associated with the ideologies of its time: speed, strength, industry, modernity, and a popular vehicle for "the people." The car, in this era, had a flair of utopianism to it. After the war, it symbolized a futuristic, systemic, and empowering world. People who drove a Ford or Cadillac in the West; or a Dacia, Trabant, Lada in the East were moving in something designed to make them feel empowered. But the empowerment—on both sides of the Iron Curtain—was only superficial. The use of cars denied society its most fundamental ingredients such as natural motion and communication. Ironically, cars would eventually deny mobility to countless thousands, because in greater numbers, the automobile alienates people from each other, as we shall see.

Isolation

Driving is the form of movement that is most contrary to human nature. Motion in a car is not a motion through space like other means discussed thus far. It is more about creating a moving space, but unlike anything of industrial scale discussed in the previous chapter. The reason for this is simply its scale. The private moving space of the automobile is inextricably linked to a sense of ownership. We are deeply entangled with them because of this sense of them belonging to us, and also because of their own technical complexity. That entanglement we have reinforces our resistance to the reality of constant motion. Although we can travel far and wide in a car, it holds closely to the post-nomadic worldview, one of objects, property, and the thingness discussed in Chapter 5.

There are several ways to see the isolation stemming from the automobile system. One of these is through the phenomenon of road rage. Research in this area points to the insulating nature of a completely artificial environment, as well as other factors regarding how it moves (blind spots, other road users etc.). Take the example of a traffic jam. It leads to psychological tension when a car's mobility is restricted. They are supposed to move, not to be stationary, and kinetic inactivity takes the control from the driver, and their command of space. This, coupled with the inability to communicate, leads to a rage.[36] Drivers find themselves boxed inside, inching along, yet sitting still, isolated, then lashing out. It's the very opposite of what our static urge points us to, which is to overcome these types of emotions.

Some more systemic examples of isolation include emergency or crisis situations when we take into account those who cannot access the automobile system. One of the tragic examples is what happened in New Orleans in 2005, when Hurricane Katrina hit the city. The biggest divisive factor affecting those from the natural disaster wasn't race, age, disability, or class. It was car ownership.[37] People without cars couldn't escape the city while at the same time, the traffic congestion was too heavy for emergency services to utilize the infrastructure. Thus, we have in the modern city a set of "mobilities," which are about moving within structures of power, social hierarchies, and giving rise to friction between certain groups. Once outside the system, one is therefore marginalized, outcast even.

There are more immediate and everyday social effects too, as shown by one study from the early 1980s. Donald Appleyard, a professor of urban design at Berkeley, did a study of social cohesion on three different San Francisco streets. He chose three very similar streets in terms of socio-politics and demographics, but with one variable: traffic. The results from his survey showed that the heavier the traffic on a given street, the less people know their neighbors; and the less they can rely on them or trust them. Appleyard also discovered that the more traffic present, the less safe people feel, and less connected to their community.[38] Only a year after his groundbreaking study was published, Appleyard himself was killed by a car in a hit-and-run accident in Greece.

Pedestrian safety is yet another example of how modern-day automobile use alienates people from each other, contributing to a less integrated society. That cars are not perfectly safe is not news, but nevertheless the numbers are shocking. Annual road deaths worldwide in the early twenty-first century amount to 1.25 million.[39] It's worth mentioning that this number is for deaths only, and does not include injuries which amount to somewhere between twenty to fifty million per year.[40] Ralph Nader, who spearheaded the first large-scale campaign for automobile safety in the mid–1960s, says that "nearly one-half of all the automobiles on the road today will eventually be involved in an injury-provoking accident."[41]

Traveling by car is by far the most dangerous way to travel. But there is, nevertheless, a strong feeling of safety. And this is the crucial point. Those inside the cars are far safer than those outside. They are the ones protected by side-impact bars, airbags, seat belts, and so on. Not only do cars not protect those outside, but there is an inverse relation between the levels of safety between those inside and outside. The safer one is inside, the less safe are those outside. The reason is because the safer a driver *feels*, the more risks they take.[42] This is a natural result of humans operating these vehicles and was named the "Peltzman Effect" after the economist who first discovered it in 1975.[43] The new safety regulations that Sam Peltzman was evaluating made the cars safer only for those inside them. Injuries and fatalities rose for cyclists and pedestrians, meaning that the danger becomes *displaced* from those in power of the vehicle to those around it.[44]

As well as externalizing the danger to the pedestrian or cyclist outside the vehicle, the automobile system as a whole increasingly externalizes the relevant responsibilities too. One example is the white reverse light, which switches on to alert those behind that the car will move backwards. Nowadays, an increasing number of models have an alarm sounding when put in reverse gear, to warn anybody who may be behind. Although it's ostensibly a safety measure, the responsibility for the actions of the driver has been involuntarily outsourced to the potential victims of an accident. The message here is that a car owner is entitled to this space, but is not responsible for their actions. It is a political discourse, where the victim is at fault, and unsurprisingly, car accidents claim Hispanic and Black lives at far higher rates than Whites' in the U.S.[45]

Space and Freedom

Cars are the kinetic equivalent of neoliberal capitalism, being deliberately inefficient and wasteful. In a gasoline-powered car, approximately 66 percent of fuel is lost completely, not translating into any motion.[46] Cars have been found to be in motion only 3–4 percent of the time.[47] They are the symptom of an economic mindset of manufactured demand. It is the model of the product creating its own market like an economic version of the *perpetuum mobile* from Chapter 5, more cars allowing fewer other options but even more cars. In the early years of their production Alfred Sloan implemented a policy of planned obsolescence, the idea being to update the models each year, to bring an element of fashion into their manufacturing, and it worked.[48] Drivers were constantly incentivized to update their models, starting a cycle of economic inefficiency and blossoming profits.

Some of the great aspects of neoliberalism are the ideals of freedom and individualism, and from the perspective of the individual, the automobile *is* a freedom machine. We can go where we want, when we want. We can travel less often to do shopping or other errands, and not have to be vulnerable to anti-social behavior on public transport. We can go places not served by public transport too.

Where the railways were fixed and inflexible, the automobile as a complete system, would provide far more reach. In a way trains fed our desire to move, but they also amplified it. The car breaks its passengers free from the inhibitions of timetables and stations. Where the train brought us into new spaces and places, the car gave us agency over these. The car empowers in a different way to other modes of kinesis. But as one urban sociologist puts it:

> Today, we experience an ease of motion unknown to any prior urban civilization, and yet motion has become the most anxiety-laden of daily activities. The anxiety comes from the fact that we take unrestricted motion of the individual to be an absolute right. The private motorcar is the logical instrument for exercising that right, and the effect on public space, especially the space of the urban street, is that the space becomes meaningless or even maddening....[49]

The problems only arise when we examine cars from a systemic point of view. They are the biggest factor to be considered in urban design. The automobile is a major cause of urban sprawl. If we think about suburbia, it's as if the space is being pushed out of the center, in perpetual marginalization, making us increasingly more reliant on cars. Where the train shrank space, the car effectively amplifies it.

Cars too often give very little for what they take, when considering space, cost, and danger. They last a very short time, when considering the

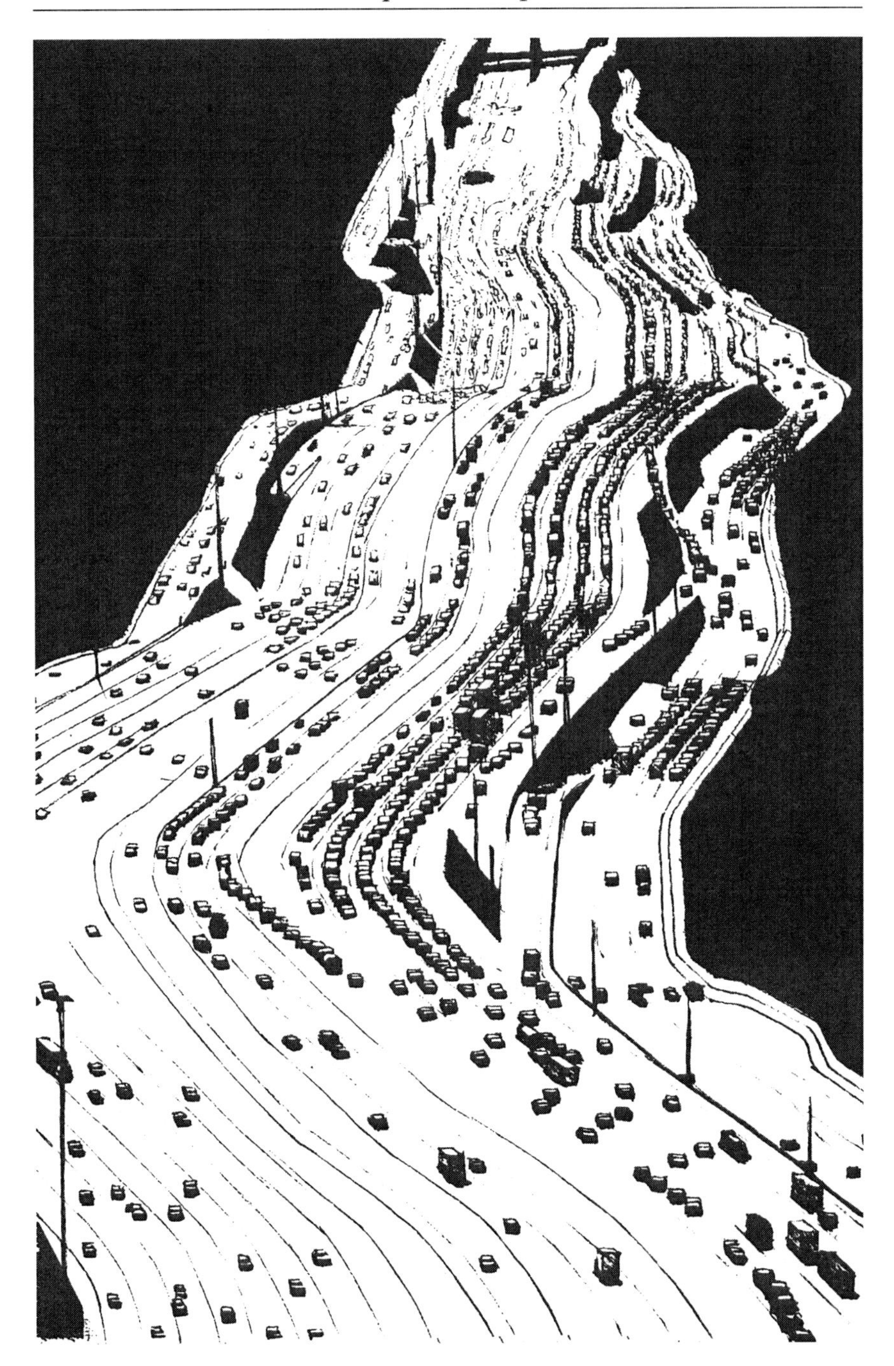

Fig. 11. Artist's aerial view of the Katy Freeway outside Houston, Texas. Spanning twenty-six lanes, it is the world's widest road.

environmental and material costs involved, such as manufacture, gasoline, maintenance, taxes, tolls etc. One study in Hawaii found the average length of car journeys is surprisingly short—just under six miles.[50] In terms of solving an engineering problem, it's not logical to shift about 4,000 lbs.—the average weight of a car—just in order to shift 155 lbs.—the average weight of a person.

All this may lead one to ask, why can't we simply build more roads to deal with the issues of space and congestion? It seems intuitive and a reasonably straightforward solution, but there's something more complex at play here, and it's where the behavior of an individual driver interacts with that of the crowd. It has been noticed in a few cases that extra roads being opened in order to deal with congestion have actually made drivers' journeys *longer*. This is because drivers have more options, and when an individual takes their favored route, it's likely that others will follow. Known as Braess's Paradox—after the German mathematician Dietrich Braess—it's a somewhat frustrating explanation of why our ever-wider roads still give rise to ever-higher use.[51]

The Now and the Future

Throughout the twentieth century, the car has gone from a rare luxury to a ubiquitous item. In order to maintain the luxury appeal, there has arrived on the market a new bigger, better version of a car—a sort of bastard child of the limousine. This is the SUV (sport utility vehicle), designed in theory for off-road use although a large percentage are driven exclusively on the road. Such models have become increasingly popular in the United States and elsewhere.[52] The greater height of SUVs is identified as the cause of many accidents.[53] From this point of view the driver sees less of what's going on in front of them, and in the event of an accident with a pedestrian, the crash point is closer to the level of the chest, which is far more dangerous than at the legs. They are the most private of vehicles, their thick doors built like the walls of a spacecraft. This cynicism towards both the urban and natural environments is demonstrated in the passengers being as removed from their environment as is possible, the advertisements bearing overtones of domination of nature (although it may be more accurate to say that it's a demonstration of a fear of nature), their astonishing fuel inefficiency, and even the availability of weaponized models, complete with pepper spray dispensers, bullet proof windows, and blinding lights.[54]

There are still many other options to explore, such as the much-anticipated self-driving models, despite claims of their better safety being mooted.[55] But because it's a ubiquitous system, which self-propels its needs, it is likely that the automobile will continue to dominate until one crucial factor supporting that system is no longer available. The car is genuinely the inhabitor of our cities, all of which will have to be redesigned in a future where we don't use them anymore.

CHAPTER 10

A Map of Consciousness

No man is an island, and even less is any city.[1]
—James Morris

We have now seen enough forms of kinesis to show that there is embedded meaning within how we move. Vehicles carry their own myths—the car is key to freedom; the bicycle an emblem of suffrage; walking and running are closer to the deeper fundamentals of what makes us human and so on. In addition to this, they express the kinetic urge to move and explore which lies deep in the human psyche.

Contrary to the kinetic urge, we have also seen how the early settled people abandoned their nomadic lives to serve some sort of spiritual purpose, expressing the static urge. Since this time, after agriculture arose and people stayed in situ as a new way of living, our motion began to be expressed through extensions of our bodies in the form of vehicles such as wagons and chariots etc. But despite settling down, we nevertheless still inhabit a world of endless motion, just like Heraclitus' river. Cities—both ancient and modern—embody both a buzzing kinesis and an attachment to place.

From a distant perspective, the city appears as something rooted and still, an expression of stasis, but that is not the experience of those within it. Cities are frenetic, energetic, and indeed stressful for many. Although it may resemble the overcoming of kinesis because of its rootedness and importance of place, it doesn't genuinely address the other side of our paradoxical equation, the desire for stasis. But what cities did do was to connect human beings to a sense of place, a bond with the ground upon which they had hunted and foraged for millennia. When the early constructions were built, for the first time there was a special quality to those particular places for human cultures.

It was with the emergence of certain spiritual developments that the demise of nomadism occurred. The relationship between place and stasis is seen at the first settlements such as the stone circle at Göbekli Tepe.

Later relations with the kinetic-static dynamic are to be found in the Viking and Egyptian ship burials discussed in Chapter 4, and later again in the wagon and chariot burials of the Yamnaya, mentioned in Chapter 6. At some level, spirituality operates in relation to the tension between the kinetic and static. It's associated with our death, with our religions, with our movement in meditative states, with the creation of societies, and even with the struggle for free mobility.

The early cultures discussed in previous chapters transformed the mere space through which we move into a thing we can identify with. Humans use place as their "raw material for the creative production of identity," says Tim Cresswell, a British geographer. "Place," he continues, "provides the conditions of possibility for creative social practice."[2]

The Agricultural Revolution cemented this association of our identity with certain places. Even today, we often like to know where a person is from when we meet them for the first time. It was the biggest paradigmatic shift in our history, from hunting, gathering, and fishing to living in settled communities with new belief systems, social fabric, and production firstly of food, but later of pottery and much more thereafter. Movement continued of course, but in a changed way. We began to move within the urban context and with an urban understanding.

Since the invention of the wheel, humans have been extending their motion through machines. Where it gets far more interesting and real is in the urban space where many types of moving intersect, because they also externalize some abstract idea of ourselves to the places we inhabit. The city is where space is divided, marked, and associated with a variety of vehicles. It's now where more than half the world's population live (a figure higher again in the developed world), and it's the place which gave rise to the many extensions of our bodies, through inventions like the wheel, the steam train, and the internal combustion engine.[3]

Spaces and Places

Urbanism caused by agriculture led to a "production of space" to borrow a phrase from the French philosopher Henri Lefebvre. The sort of space he talks about is something produced by societies rather than individuals. Considering it just by itself "is an empty abstraction," he claims.[4] Historically, from agriculture onwards, space is something measured, imagined, produced, and manipulated by the communities who share it. And what is created must meet the expectations that we lay upon them. Sometimes it is a social space, and other times it is close to being an empty abstraction. Other times, it is strangely both, as we shall see.

Many of the early settled societies that were successful have been considered as "hydraulic" by historians. This refers to the irrigation systems that were built and used in places like Mesopotamia, China, and India. This wasn't simply a canal from a river serving water needs of a town, but sometimes a vast array of water lines organized from the highest political level to meet the demands of agriculture. Without this infrastructure they could not have built such complex cities and societies. In terms of space, this is one of the embedded ways of striating it, cutting lines into the ground, dividing and geometrizing the terrain. It was the earliest complex societies which did this not long after the great settling, and not long before the invention of the wheel.

It is the striation of space which allows societies to exist at a far more complex level than that of hunter-gatherer groups. Roads, lines, rules, and hierarchical power structures all depend upon delineated space.[5] The nomadic peoples wouldn't have had these relations with it, and even some empires have existed that have not dealt with space in this manner. You may remember from Chapter 6 that I discussed the Mongol empire and its vast area conquered from horseback. One of the elements of Genghis Khan's territory was that it maintained the "smooth" space of the open steppes, with no construction of roads, without having to embed a huge network of lines, other than the relatively intangible ones as the postal system.[6] It took the form of a single texture. It was the same for the early Pacific sailors, traversing the smooth space of the ocean, interacting with it in a receptive way, smelling the reefs, and observing the birds.

The first cities were sacred places. But soon enough the striation of space began to carve up the land as it became owned, and food hoarded to mitigate against poor or diseased crops. Ownership of objects and our entanglement with them quickly infused our way of living. The static urge involved with those spiritual revolutions and early settlements became dominated by our entanglements. Early civilizations quickly became thing entangled, associating themselves with specific places. If all of this sounds familiar, it's because we are continuing to live this way today.

If you remember from Chapter 1, Heraclitus' river depicted endless motion as the reality in which we live. One cannot step into the same river twice, he tells us. One of the ways to respond to this truth was through the kinetic urge. Exploring, expanding territory, engaging in a meditative run are methods of plugging in to that reality, uniting with the flow of life. But there was also what Cratylus said of the river, that because it's always in motion, and having no essential thingness, it cannot truly exist. The relationship between Cratylus' statement and stasis is far more mysterious and intricate.

The irony is that although the static urge may have driven the first spiritual settlements to exist, the result was the rise in belief of a thingness as those cultures became increasingly entangled, technological, and materialistic. Agriculture and settled living rejected the truth that Cratylus was pointing towards. Turning away from seeing the world in that way, early settlers instead built a world where measuring and delineating space and owning products went hand in hand with the kinesis of our bodies extending into other forms of movement, and gradually less to do with the initial spiritual stasis. It is no coincidence that one of the most striking features of these early societies was the remarkably high level of knowledge of geometry.

The City of Consciousness

The city emerged from the tension of kinesis and stasis, but the static end of the equation was offset by this involvement in the mishmash of objects. The search for stasis was put on the shelf and cities only made our restlessness more apparent. What is a city? we may ask. A city is the only place where all the various species of kinesis collide. They meet here—for better or for worse—in the theater of movement and its politics, where we watch the ramifications play out. A labyrinth that holds a variety of kinetic modes, motion not only occurs here, but it dictates how the space around it must behave. The city is the spatial stillness built around the needs and demands of the kinesis that contains it. At the same time, it is the embodiment of humanity's abandonment of nomadism. The motion of a city strangely represents the closest anthropological phenomenon to stillness that humans can achieve—the sedentary lifestyle.

The city as a whole is a relatively still thing, or at least it's probably not uncommon to consider it that way. But simply look *within* it to see that it is positively buzzing with motion. The kinetic expression of our nomadic ancestors was turned inwards to the city. In reflection of this, human consciousness began to resemble itself in the city, each dovetailed into the other. Rebecca Solnit, a historian of walking among other things, relates the city to the phenomenon of modern consciousness, in her typically elegant way:

> A city is built to resemble a conscious mind, a network that can calculate, administrate, manufacture. Ruins become the unconscious of a city, its memory, unknown, darkness, lost lands, and in this truly bring it to life. With ruins a city springs free of its plans into something as intricate as life, something that can be explored but perhaps not mapped.[7]

Here it is, a physical counterpart to a human consciousness. It's not by deliberate act, nor is it entirely a coincidence, but somewhere between the two, vaguely suggestive. In the same way that an Inca stone wall resembles the shape of the corn on its cob that its builders ate, the cities of humans resemble their consciousness. Our consciousness is so many things at once, contradictions, connections, tendencies, and habits; reflected in multiple industries, politics, striated space, transport, markets, and bustle of the city. The mind is intangible certainly, but not static. If we directly experience it, it's always moving.

The chaos of old memories and meaning stretch across in layers under our feet. Through new building, regeneration, decay, and even gentrification the city acts as an archive of memory.[8] There is so much of the past all around us, hardened into the walls and ground, but also there is future development always on the horizon. The city is a record, albeit in tatters, of the previous cultures that have existed there. Just like consciousness, it is comprised of memory. Just as the railroad shrunk space, so the city compresses time.

Like many mental compartmentalizations, the city organizes its space geometrically. Walls, lines, parks, and tunnels form a warren of spaces that mirror our routines of navigation through our inner mysteries. For centuries, foot traffic was the only such movement allowed within the European city walls. Comprised entirely of small streets, the medieval city in Europe was typically undesigned and chaotic compared to some of its counterparts.[9] Once the Renaissance blossomed, space became both public and architectural at once, an arrested movement, frozen in time, and powered by geometry.[10]

Vehicles in Cities

One of the keys to the city being like this was that wheeled vehicles were normally kept out as much as possible since ancient times. In the sixteenth century, when better steering on wagons made them more useful inside the city gates, they were brought into the centers, but were strongly resisted by many.[11] They were dangerous, took up space, and forced pedestrians to the sides of the walkways. Unfortunately for them, the new vehicles were here to stay, and it was having this extended motion inside the walls which kicked off the design of more modern cities. One of the results was that new spaces were designed with wheeled traffic in mind. This was manifested in the completely straight and broad avenue, notwithstanding the irony of it being down to new advances in steering. Nevertheless, it was the straight line which became a new geometrical dominating force for city planning in Europe.

With it came homogenous facades, straight horizontal lines, and vanishing points in the surrounding architecture. Despite the majority of those using or living in the Baroque city being pedestrians, the large-scale attitude of the design was to accommodate the few who went about in horse-drawn vehicles. The avenue is not merely a large-scale street, the visual design is completely different. The *psychogeography* (the mental effects of one's environment) of the avenue to those in a carriage is a world apart from that of the pedestrian.

Walkers enjoy a visual variety at their speed, being able to focus without effort on what surrounds them. The speedy traveler, on the other hand, enjoys repetition. Looking from the inside of a carriage, close details are blurred, and regular rhythm becomes important, hence most boulevards are lined with trees at equal intervals. They create a pleasant and soothing repetition to the mover's eye and mind. The stillness held in that recurring image of identically shaped trees, counterbalances the motion of the horse and carriage.[12] This design was the antecedent to the experience of cinema from the train window, discussed in Chapter 8.

Through its discourse of power and speed of the vehicles which drive on it, the avenue dictates how the city is to be. It was the wagons and later the chariots which were the instruments responsible for geometrizing space, not the architecture. The avenue is what channels the city's motion, making for a city which is designed for speed, by a straight-line kinesis. With increased speed comes a new social power, realized through motion, and whoever it is that moves fastest is the one who dominates.[13]

The increasingly powerful elite class were now in carriages or on horseback, moving far faster than the regular citizens, and conquered space through the threat of injury. The rich no longer had to wait for poor people to get out of the way, they were all pushed towards the gutter, quite literally marginalized by the street design itself. Socially speaking, the kinetic effect of the avenue was huge, leading Mumford to state that "The avenue is the most important symbol and the main fact about the baroque city."[14]

A theme of kinesis became apparent, not only through the allowance of faster vehicles, or the visual repetition, but also by geometrical invention. The regularity and repetition of building facades, and the new emphasis on horizontal features on them led the eye towards the vanishing points (this is where parallel lines appear to meet in the far-off distance), a style which emphasized motion, speed, a far-off destiny even. Although frozen, its kinesis was apparent through many new tricks such as rhythmic sequences and brief interruptions on the horizontal lines.[15] The glory in secular architecture is realized at the vanishing point, to which everything is directed by the devices that the city planner had at hand.

The grandiose quality of the accompanying buildings in the Baroque European city, was reflected wonderfully in the redevelopment of Paris of

Fig. 12. An artist's sketch of a typical Baroque-style street. Note the relation between the building facades and the kinetic thrust of the street towards the vanishing point, located at the center.

1853–1870. The potential for military use of avenues explains why many of these are so oversized. They were not merely built for horse and carriages, but for Napoleon III's armies, and they are effectively reminiscent of them in their absence. Displacing about one million of the poorest in that city at the time, the planner, Baron Haussmann famously initiated the first truly modern city. Far longer than the Baroque avenue, the Parisian boulevard is a straight line modeled on the railway.[16]

Desire Lines

The rectilinear qualities of city geometry create a certain inorganic quality in the urban fabric. To the pedestrian it is hostile and awkward.[17] People don't naturally walk in perfectly straight lines only to take a hard ninety-degree turn. They generally follow trodden paths, or sometimes create new ones, which lead in the most natural and efficient manner to a given destination. The crisscross of streets and the straight-lined avenues are for wheeled traffic only, and naturally discourage the pedestrian.

This fact was incorporated into the work of one American architect, Christopher Alexander. Proposing the idea that the natural method surely ought to better the designed product, he planted grass in the areas between some buildings, and left pedestrians to trod paths themselves, only later to pave them.[18] His experiment was a celebrated success, both pragmatically and aesthetically. The lines that we see left from repeated previous pedestrian journeys are called desire lines and highlight the difference between what's usually provided and what pedestrians are organically drawn towards as they move. Desire lines are especially prominent in suburbia, a fact which tells us how the car saturates suburban design.

Each step contributes a footprint, all of which accumulate to form a desire line. They record the previous treading by unknown walkers before us, compacted over time to form a path deviant from the ordered plan. It may cut through a corner, or lead one through the woods. They are noticeable for straggling away from the square and rectilinear built environment as distinct lines peel off from the cement footpath. They aren't designed, but arise in an organic fashion, and some geographers have even mapped them out as alternative city geographies.[19] We're nudged to follow them, taking confidence by following past walkers, connecting with them through our own steps.

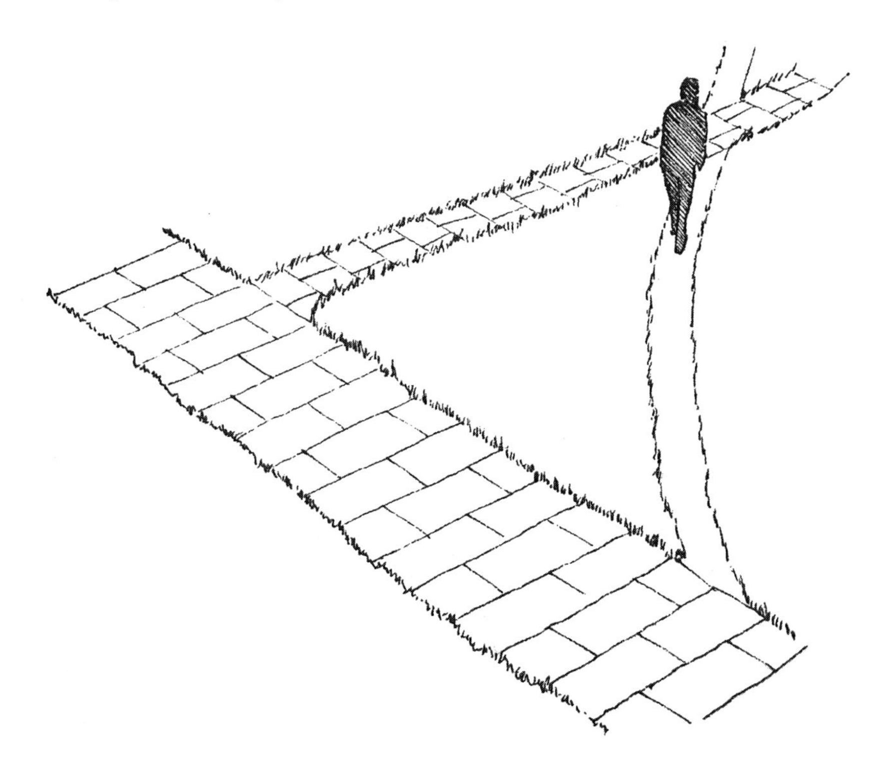

Fig. 13. An example of a desire line.

The suburban world is born of the technocracy of automobile infrastructure, but also of a mix between the static urge and our materialistic tendencies. For many, there is a strong desire to live in one's own home. But sadly, it necessitates a world of car parks. The main and often the only means of transport for suburbanites is the car, and to afford ease of movement and fluidity with parking for each car, one needs somewhere between three and eight parking spaces.[20] The reason for this is that cars are only in motion about 3–4 percent of the time; the rest of the time we don't use them.[21]

One remarkable fact about pedestrians in the modern world was noticed by the celebrated architecture critic and activist Jane Jacobs. In Disneyland, she observed that although it was a pedestrianized zone, people nevertheless used only the sidewalks.[22] It may strike one as an inconsequential example, but what it can tell us about the modern pedestrian is quite insightful. Even when traffic is not there, it's there, continuing to exist in the minds of those who are for once able to move freely within the built environment. The walkers on the sidewalks in Disneyland demonstrate a total alienation in movement from the reality of the space that surrounds them. Ultimately, sidewalks are an abstraction that serve to keep the road clear for cars more than they do to satisfy pedestrians' needs.

As one of the industry experts tells us, "cars have many external costs, but the cost of parking in cities may be far greater than all these other external costs combined."[23] We are increasingly surrounded by multi-story car parks and drive-in food outlets, and parking costs two-thirds of the total shopping mall construction in the USA.[24] Urban space is becoming mechanized to be part of the greater automobile system, and many of us have no choice but to inhabit it in this way. The result is that people are increasingly disconnected from their cities, one piece of research claiming that 75 percent of people feel no attachment to where they live.[25]

Non-Places

As the world of car-parks grows, new urban spaces are increasingly appearing that have a tone of blandness to them, despite their clean and orderly appearances. And after they are built it's remarkably difficult to do anything about it. This ruthless lack of spirit is a relatively new phenomenon, and the results have been described by Marc Augé as "non-places."[26] They also include airports, waiting rooms, and shopping centers, basically anywhere that a person can feel so disconnected from the place they're in, that they or the place itself are interchangeable. These are the spaces through which more and more of us must move as the world's population continues to urbanize.

These entities are a type of social space mixed with being the empty abstraction that Lefebvre referred to when describing space understood in isolation. Non-places, we can say, are a form of placelessness. They prove to us that space is no longer curated, but simply bought and used. During the European Renaissance, architectural, urban, and public space were all one and the same. It was a coherent single space—visible when one walks about in a well-preserved city from that era today. There is a uniform concept at work in these places, reliant on the beauty of the buildings and their materials. The last bastion of spirituality was that the common height of these builds had to be less than the nearest church. There are some secular examples of this from more recent history, such as Le Corbusier's planned city of Chandigarh in India, where a sense of uniform coherence binds everything together, and it's designed to be best experienced as one moves through it.

Non-places differ in their uniformity, because they have banality at heart. It is divorced from any aesthetic concept underpinning it, only utilitarianism. Buildings are often designed without any sympathy to or awareness of the neighboring ones, another striking discourse of hyper-individualism. Ironically, the result is a whole ensemble of bland spaces, oozing into each other, such as office blocks, housing estates, and highways. It sucks the meaning out of common spaces, because they tend to be designed only to perform a utilitarian function. We must make do with the character-lacking open public spaces of the modern city. We wait for the metro, the airplane, or sit on a plastic mass-produced bus shelter, the same the world over. It's no accident that a modern conception of space is a *lacking of things*. It is empty, homogenous, and divisible, rather than being a thing in itself. A natural landscape, on the other hand, lacks nothing. It is an entirely positive space, being something in and of itself, with its qualities and characteristics. But many of the public-use modern city furnishings are simply nowheres.

But we could think of place not as a separate thing to ourselves, but as the bonding agent of all moving beings. Space is part of the global flow of things and people, being dynamic, relational, and open.[27] Perhaps there is more to it than just yet another part of our environment to react to. Understanding space means understanding that the world *is* flux, as the great Heraclitus told us in Chapter 1.

Through their soullessness, non-places create nomads because there is nothing in them to connect with. They are places that people merely pass through, as it's transport systems which create them. There are the few nomads who have traveled by air, train, and car. And then there are the others who form another tier, who will be the climate nomads of the future, paying the externalized costs of all the fossil fuels that powered industrialization and transport.

Each city is a little different in spite of the powers of globalization and depending on how the local culture is. Motion within cities is through a produced and unnatural space.[28] It is motion strictly conditioned by rules of the road, timetables, and metro maps. But despite the differences from city to city, the movement is almost exactly the same. Once a person has learned how to use a metro system in Paris, then they are able to use one in Tokyo, New York, or Prague—useful, but nevertheless uncanny. The movement and the experience of it is the same, the passenger is alone in a crowd where communication is taboo. Trains bring people together, as mentioned in Chapter 9, but at the same time can alienate us.

This has been the process for millennia. Consider the Roman roads, uniform throughout the empire, never adhering to the natural course of the topography.[29] Roads are non-places, as they are characterless and non-descript, existing solely to connect real places. Given that this process is so long-standing, it's reasonable to conclude that it's somehow *inherent* to the city for places to gradually metamorphize into non-places.

Desanctification

One of the biggest factors that steered us off the course to connecting with stasis was the relation we developed with objects, particularly regarding ownership. And this began with our attachment to place. Then ownership of things developed, entanglement with them, and for many, dependence upon them. Living entangled with objects in a world where enlargement of capital champions many other values, is to live in denial of the nature of reality as described by Heraclitus and Cratylus. We resist seeing the river as a process in constant flow and we resist seeing the emptiness of it. Instead, our minds are geared towards viewing it as a "thing," probably because we see all phenomena this way. The roots of this mindset are in the city.

One of the proofs of this is in the *desanctification* of urban places over time. There is poignant difference between a modern-day city in the developed world and the sort of place that Göbekli Tepe was, where nomads gathered frequently for religious ceremonies. They were at the cusp of abandoning the way of life that had driven the species since its emergence, instead to live there and find a new way to make it all work. This was seemingly so important that it was done at great expense to their health, exposing themselves to plague, disease, bad harvests, and complex social hierarchies to serve their need to achieve stasis. In Europe, the cost of this was such that their poorer nutrition caused a decrease in average height which was not recovered until the turn of the twentieth century.[30]

Over the many centuries that people have built and inhabited cities, there has been a continual lessening of the sacred nature of those spaces, when compared with the earliest cities.[31] They started as spiritual places in their infancy, with worship as the fundamental reason for their existence. Today it is commerce, a simple interaction of equally-valued commodities, be that labor, products, or services. This is not to ignore the many city communities of shared interests, religions and so on, but commerce is now what the city serves primarily to sense. And we see the utilitarian nature of this coming through aesthetically in many modern builds throughout the world.

Another ramification, if not a proof, is the technologizing of our kinesis with the extension into machines. Our movement was no longer embodied in the immediate human way of walking and running. Once agriculture led to the wheel, this spurred on to so many other inventions causing ever deeper entanglements, that it's hard to imagine not living like this. Of course, they come with a plethora of benefits too.

In his marvelous book, *The Poetics of Space*, the philosopher Gaston Bachelard tells us that the home is "our first universe."[32] As we get older and look back at places where we've lived before, we see them as static, he says. Our particular memories of home are different to any other places we remember, we relate to these not as if we are historians, but poets.[33] This is closer to the genuine stasis that calls to us. In memory or imagination, places are static and also warm, satisfied, free of wanting, perhaps emotional, and sound a deep longing within us. It is to the relationship that cities have with the mind, consciousness, memory, and invention where we go next.

Chapter 11

Drifting Through Utopias

Mind takes form in the city; and in turn, urban forms condition mind ... the city records the attitude of a culture and an epoch to the fundamental facts of existence.[1]

—Lewis Mumford

Cities shape our perception, and have been doing so since the advent of the Agricultural Revolution. We perceive the reality of Heraclitus' river in conceptual terms only, as a simple piece of information, without fully absorbing the ramifications of it. And that of Cratylus? Well, we probably understand it less because it is in direct opposition to our way of interacting with a world of objects, manufacture, exchange, and ownership. Everything moves, but we understand ourselves to inhabit a world of moving "things." In short, our material outlook on the world denies us the ability to see the world in the way that Cratylus suggests, that because everything is always in motion, there's no thingness to any of it. Instead, we have become more and more thing-entangled over time.

Since then, human expression of kinesis has been through vehicles of increasing speed, power, complexity, and reliance on fossil fuels moving ever further from the nomadic-era ways—walking and running. The history of our motion points us towards how our consciousness has changed through the places we've been living in, and the types of lives being lived. The more entangled humans become, the more distant we are from the static urge. At the same time, increasing entanglement also leads to better scientific understanding and an improving quality of life.

There are, of course, those counter cultures who push against the current. This chapter is about them. It's about those who act in a deliberately useless way, rejecting the ideal of utilitarianism in capitalism, who travel without destination and who dream up places that cannot really exist. Instead, they engage with motion and the city in a way which explores the nature of consciousness, in a re-orientation towards stasis.

Enter the Flâneur

It was in Paris, soon after the Haussmann redevelopment of the mid-nineteenth century that a new type of engagement with this modernized city sprang up. Probably it was because Haussmann's project got rid of the city's many arcades, the arched coves which contained markets. Within these, people had been less kinetically active, and after they were gone, no longer having the spaces to mill about, Parisians wandered out into the world of straight-lined streets and avenues.[2] Places that were agreeable to spend some time simply peoplewatching were fewer after the modernization of Paris. And so, the people watchers were put on the march, made to move about the city as dreamy meanderers.

These urban wanderers aimlessly drifted through their massive city. Soaking up the street life, they watched the interactions and all the movement that occurred around them, people flitting about in horse-drawn carriages, trams, and bicycles. Wandering by foot was crucial to this, as imagined by Charles Baudelaire and later, Walter Benjamin. They were captured by the movement of others, and themselves moved like urban nomads. A rootless mingling with the citizenry of the modern city of Paris was an end in itself. The term *flâneur* (from Old Norse: to wander without purpose) was used to describe this person wandering about the streets, avenues, and markets of the city, as they would delight in the motion they saw.[3] The traffic and bustle were their entertainment. As Baudelaire defined it,

> The crowd is his element, as the air is that of birds and water of fishes. His passion and his profession are to become one flesh with the crowd. For the perfect *flâneur*, for the passionate spectator, it is an immense joy to set up house in the heart of the multitude, amid the ebb and flow of movement, in the midst of the fugitive and the infinite. To be away from home and yet to feel oneself everywhere at home....[4]

There is also a sense of receptivity, detachment, and placelessness to this character in Baudelaire's description. Flâneuring is a union of participation in kinesis and consciousness, a form of mindful urban walking.

Along with the importance of walking are two other important criteria. Firstly, is the concept of *dérive*. This translates best as "drift." In later reincarnations, it relates to the French Marxist philosopher Guy Debord's *Theory of the Dérive* (1956). This relates to the purposelessness of wandering in a city, the very place where people are often seen rushing to and fro, with little or no time to stop and chat, or absorb the experience. *Dérive* contradicts this, by being an action without direct reason. Urban areas are best comprehended in mindful experience, away from the turning of wheels. The city then becomes a dreamscape experienced by the wanderer.

In flâneuring, they relinquish control over destination and allow the shape of the built space around them to dictate how and to where they move.

Dérive is a pointless, unplanned wandering, with echoes of nomadism embedded in it. It is about leaving behind the usual manner in which we travel within a city. The goallessness to drifting is of central importance to this. We can bring our instinctual mind to the environment that we find ourselves in, in spite of it being boringly familiar. As the city is a strongly utilitarian entity, where all motion is justified, *dérive* subverts this idea. The *flâneur* remains potentially active as an agent in a capitalistic society, but chooses not to act, disengaging themselves from the capitalist tendency.[5]

The second criterion is the act of taking it all in in a self-reflective way. One notices the effects of moving through the urban space, and how the city bears itself within the consciousness. It is the moving city, being brought to the halt of mind. The urban wanderer operates with the role of an anonymous receptivity. A later development of *flâneurism* is its association with an artistic and political movement known as the Lettrist International, who had Debord as one of its more prominent members. Because it's essentially an anti-consumeristic activity, it lends itself well to politics critical of consumerism and appearance-obsessed modernity.[6] In this mode, it became known as psychogeography and here is where the union of kinesis, consciousness, and the urban environment intersect.

Psychogeography

The psychogeographer often explores the wasted empty spaces of the built environment, such as abandoned warehouses, empty buildings, and underground catacombs. Following Debord's worries about the desacrification of the city and its replacement by the "pseudo-sacred" image-obsessed modern aesthetic, psychogeographers seek out the forgotten parts of our everyday surroundings.[7] Traipsing over the layers of history under their feet, they also discover the ideologies that drove previous urban planning decisions or the associated political discourse which pervades the space. It is a spatial practice, but also a mindful one, because the point at which one orients themselves towards is where the consciousness is affected by the immediate environment. Walking a city in this way articulates the spaces, lines, and intervals, bringing them all to life in the walker's mind. There, they form a map, maybe not unlike Tim Robinson's maps from Chapter 2, which were "dreamed in footprints."

Regardless of if we refer to it as flâneuring or psychogeography (some even prefer to call it "deep topography"), this activity is one with the constant movement of consciousness in its focus. Psychogeographers, we could say, are *urban kinetic meditators.* Observant of what goes on mentally, they use only the body to move, in a harkening back to our nomadic ancestors. They navigate without destination, having freed themselves of the tyranny of the clock, in search of meaning via kinesis. There is a glimpse of stasis in this type of activity, because of the element of observation of kinesis.

In this tradition, we find an alternative meaning of the city and what it might mean to be in it. The philosopher and sociologist Michel de Certeau explains,

> To walk is to lack a place. It is the indefinite process of being absent.... The moving about that the city ... makes the city itself an immense social experience of lacking a place—an experience that ... create[s] an urban fabric, and placed under the sign of what ought to be, ultimately, the place but is only a name, the City.[8]

This practice of psychogeography is an inner-oriented pilgrimage, through the desacrified non-places that surround us in today's environment. A totality of placelessness surrounds the nomadic pilgrim in one great non-place. It is a search for "what ought to be," lost in the lack of spirituality, and using what remains of it to guide the practitioner, who rambles over the archive of compressed time that is the city.[9] By walking through the non-places, they attempt to put meaning back into those areas.[10]

There is another, much older practice of uniting mind with place, one where stasis is achieved, we could say. It involves both the non-place and the unmoving, and it can be included in the rejection of utilitarianism that I mentioned at the outset to this chapter. It is the utopia, the non-place *par excellence.*

Our relationship with place, space, and mind, as well as the kinetic and static urges, all come together here, in this unreal and impossible city. Going there is traveling mentally, but without any real destination, it is also a useless city, having no physicality whatsoever. But nevertheless, human minds have been creating utopias for as long as we care to remember. Why? It is a manifestation of the static urge. New cities which exist only in the mind, free of motion, are an attempt to get closer to the perfect stillness that is sought after by every human. We strive for the perfect society, just as we strive to be delivered from kinesis, from endless restless moving. It's as if the destiny of the *Homo sapiens* is a final resting, not in death, but realized in a utopia.

Geometry

In Chapter 10, I discussed the geometrizing and striation of space, something that has been part of city culture and of almost all civilizations that we know of. I mentioned how it may have grown from the hydraulic nature of those societies, in their irrigation systems for agriculture. Geometry is a further development of settled culture, and here is where utopias existed in embryonic form. The geometrical world is one where there are no imperfections. Its straight lines really are perfectly straight, the circle is also perfect, and lines have no thickness. It is objective and unaffected by human actions. For example, if the world were to end tomorrow, the square of the hypotenuse would still be equal to the sum of the squares of the other two sides. This theorem—first discovered in Ur but traditionally attributed to the later thinker, Pythagoras—exists over and beyond the human world. Although it's a discovery made by the human mind, its existence lies outside of it. And we access it through the imagination. The field of geometry was one where those who participated in it could exercise their mental movements in a secondary, abstract, theoretical world. It resembles the carving up of natural space that so many cultures have done through irrigation, streets, and territories.

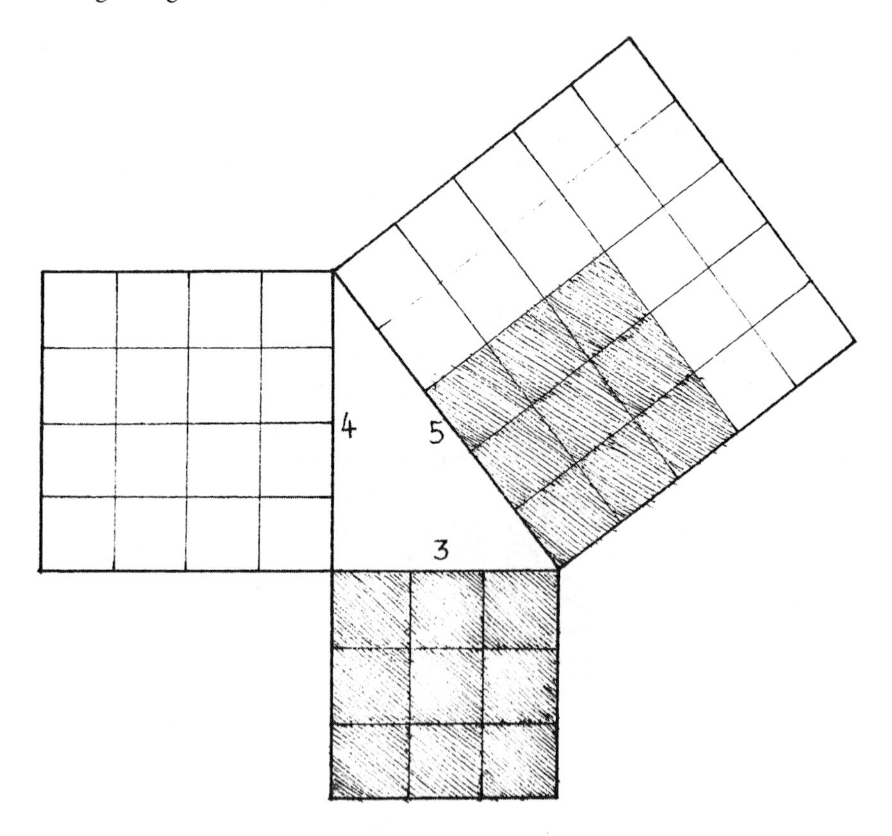

Fig. 14. The Pythagorean theorem, $3^2+4^2=5^2$ or 9+16=25.

It is no accident that the earliest occurrence of the Pythagorean Theorem was found in the remains of one of the oldest cities on Earth. What we see here is the development of nothing less than a new type of human consciousness. Historically this unfolding ran in parallel with the rise of civilizations—urban bound, agricultural societies. Ever so slowly, the great changes that were happening in most human populations were affecting not only the way we think consciously, but the whole layout of the mind itself.

These changes that arose with the city are manifest in a highly rational consciousness first seen in the fifth century BCE. If we had to pin the emergence of a new consciousness down to a single moment, it would not be the dawn of mathematics, but understanding of space through geometry in Greece. Why? It is because it demonstrates the creation of an abstract world, existing wholly in the mind of mankind. It was the application of mathematical thinking to space, a rational manner of understanding that which we move through.[11]

Of course, the geometrical world is a static one, simply because it exists only in the human imagination. But it has its roots in kinesis. We already know that the perfectly straight line—or perfect circle—is not commonly found in nature, as famously said by the Modernista architect Antoni Gaudí.[12] But once we bring motion into the equation, it *can* be found. Light travels in straight lines. Objects fall in straight lines. In Chapter 5, we discussed the circle in nature, and how it may have come from the human mind observing rotation of the Earth, and how that related to the wheel. If you throw something, a ball say, it falls in a parabola. Parabolas are also observed in water fountains. It's interesting that one of Gaudí's most common structural forms used is actually a parabolic arch. Once movement is involved, there are geometrical forms in nature, plenty of them in fact.

The geometrical world can be thought of as a form of utopia because of the perfection in its forms. In both geometry and utopianism, we come one step closer to resolving the paradox of the two opposing urges of kinesis and stasis, as it is in these mental realms where the two are deeply interrelated.

Utopias

The utopia (or dystopia) is a static world. They are most presented as places where all problems have been resolved, they are perfect, and stay that way. In the last chapter, Gaston Bachelard reminded us that we see the past places we've lived in as static. It's not simply the building or space which is unmoving, but the situation. It's as if we are looking at only a single frame of Muybridge's set of photos, a static moment, temporally frozen. Time is static in utopias too. There may be movement depicted but never any real change. In fact, once some sort of change happens, the utopia often falls apart.

The word utopia, coined by Thomas More in his famous book of the same name, published in 1516, comes from the Greek *topos*, meaning "place." By adding the prefix "u," he negated it, meaning "no-place," but it also can have the meaning "good place." So, More's term refers to a good place that doesn't exist.[13] Utopias release us from the reality of our problems that we face in daily life, providing an escapism to a perfect world. They may not physically exist, but being non-existent and imaginary are not exactly the same thing. Just like the value of a dollar, its existence in the common imagination does not preclude its existence entirely. Utopias have a value, a meaning. They can be a means of working out a better society from the one we have and a way of understanding why things are how they are. In what is an ironic historical coincidence, Thomas More's static world in *Utopia* is modeled on the philosophy of probably the first known thinker to really examine the nature of motion, Epicurus, and more will be said of this interesting thinker in the next chapter.[14]

Although we could consider the Polynesian sailors of Chapter 5 to have engaged in utopian acts (as they went with the intention to settle new islands), more often than not, utopias have been cities. The abandonment of nomadism was also a utopian act, establishing a new way of life, ordered by geometry and spirituality, by ideas. It was an attempt at addressing the longing for stasis. Strangely, we see from the earliest examples, such as the garden of Eden, the Australian Dreamtime (discussed in Chapter 2), or Plato's *Republic* that there is a strong tradition of them having been in our *past*.[15] Why might this be? It is because our utopias never worked out so well as we may have imagined—hence More's superbly accurate and poignant term for them. Utopias show us the paradise that existed before our nomadic ancestors settled.[16]

Many have a similar story bridging the gap between then and now. All had been going splendidly, until some sort of fall from grace happened and that explains how we ended up in whatever mess we're currently in. The distant and now long-forgotten past is a place of great natural beauty, perfect weather, and an abundance of food. It was our kinetic side which caused the problems. Adam's curiosity drives him to taste the apple. This traditional story seems to be in so many cultures that it points to some sort of pan-human experience. These tales, and nostalgia for a lost past, recall a time where there was a better quality of living.

The utopian ideal is one where the duality of kinesis and stasis doesn't feature. Those living in paradise do not have an itch to scratch, or a dissatisfaction to attend to. They are whole, always content, and never bored. Obviously, this is too far-fetched to have any semblance in reality, but current research points towards utopias being rooted in the nomadic, placeless life of our ancestors.[17]

Some aspects of nomadic life do indeed match the typical utopian model. The hunting-gathering lifestyle supported a much easier life when compared even to today's life for many people in terms of work. Societies were far more egalitarian, with food often within reach.[18] Theirs was a society with little hierarchy, where all basic needs were met, the world of work required fewer daily hours than people generally work today.[19]

Utopias demonstrate the existential restlessness of the post-nomadic lifestyle, an agitation that occurs deeply within. Settled life simply does not satisfy our static desire. The proof of this is in the very existence of the utopia as an imaginary human creation across so many cultures. It attempts to quench that thirst for *home*.

Second-Order Thinking

From the times of the Sumerians, through to our friends Heraclitus and Cratylus whom we met in Chapter 1, there has been a tradition not only of building great cities but of dividing space itself. The field of geometry, greatly expounded by the Pythagorean school of ancient times, did exactly this. By dividing space, understanding it in a new technical way, they conquered it, in a way domesticating it from wildness. Geometry represents the first endeavor in purely rational thought.

Geometry depends on movement for its lines, curves, and shapes to exist. Like the spurting of a water fountain describing the arch of a parabola, it's born from motion. The philosophical team of Gilles Deleuze and Félix Guattari—when writing about nomadism—mention how the striation of space comes from movement, "[Space] is striated by the fall of bodies, the verticals of gravity, the distribution of matter into parallel layers." This geometrical approach towards space "formed an independent dimension capable of spreading everywhere, of formalizing all the other dimensions, of striating all of space in all of its directions, so as to render it homogeneous."[20]

But because geometry exists in abstract space it also depends upon stasis. It occupies a similar space to a utopia, non-physical, idealistic, formal, and without imperfections. All the same, what knowledge we gain from geometry—and also the political knowledge we may gain from utopian thinking—is nevertheless applicable in the real physical world. It is no accident that this is also the time when the first maps appeared. The bird's-eye view, looking down as if from elsewhere, is distinctly abstract. The object of study is the world we are familiar with, but the viewpoint is otherworldly. That abstract space relates also to Plato's famous world of forms, where the idealized, essential, and perfect versions of things exist in some transcendent manner, and we have the rough copies of them in the world that we inhabit. It is no accident that Plato's Academy bore the inscription above the door saying that "let no-one unversed in geometry enter here," or words to that effect, such was the importance of the discipline to thought.[21]

In its time, this was a new form of thinking, and is now referred to as *second-order thinking*. It's not a linear progression from what preceded it, as in a "better" or "more developed" level of understanding. Second-order is simply a different approach, and is most easily defined as "thinking about thinking."[22] This is about focusing on the mind itself, and questioning its nature. Geometry parallels it, in that it relies upon proofs of its theorems, just as rational thought can be defended by logical argument. This is the moment when the first utopias were conceived of; they emerged from this new type of thinking, as if utopias were sharing the same abstract space with the perfect lines and shapes of geometry.[23]

It was not simply a new method of rational thinking that sprang up which led to utopianism and geometry. These were symptoms of a far greater set of changes. Some thousands of years after the Neolithic Revolution, Plato and his contemporaries demonstrated a new form of consciousness altogether, one with a desire to understand kinesis fully, and to overcome it spiritually. It is the part in our exploratory story where motion and mind combine in grasping at transcendence, at spirituality, even at the impossibility of a utopia. Second-order thinking relates us to perhaps the deepest aspect of what it means to be human, to get to a resting point, to stop moving.

The reaction against the post-nomadic lifestyle came in the form of a spiritual revolution which attempted to quench the thirst of the primal nomad living inside of us. The agriculturalists brought about what we now understand to be capitalism—an overproduction of food and hoarding of other goods which was matched by explosions of culture and population. But the human itch to move, to explore, to conquer was to be faced and examined directly in this epoch which is understood by some as the most developed time in human spirituality.[24]

This was not simply a cultural phenomenon arising out of ancient Greece. Just like the Agricultural Revolution that occurred in many places at roughly the same time, this revolution of consciousness—if we can call it that—also happened in several places at once. New movements of spirituality took hold in Greece, Palestine, Persia, India, and China. People started to spiritualize, and leave the cities for a simpler life.[25] Initially it was those who had been marginalized by the city society, slaves and foreigners, centering themselves around cults to the likes of Greek gods Bacchus and Orpheus. It wasn't merely a return to a simple agricultural life, but a kick back against the value system of power structure, class division, and the structural violence of hierarchy. Lewis Mumford describes this as a "profound revolt against civilization itself."[26]

The Axial Age

This spiritual revolution was first described by the German philosopher Karl Jaspers, and he coined it the "Axial Age," based on the idea that this was the axis of human history around which the most important spiritual developments turned.[27] As to why this happened, he suggested that we lend credence to an idea that came from the economic historian Alfred Weber. Weber maintained that the use of horses opened up a new understanding of space. It was riding on them that gave a completely new experience, he thought, and the spatial vastness that people dominated had a direct and profound impact on their consciousness.[28] It must be noted, however, that this idea has been criticized by some as an absurd oversimplification of a far more complex and subtle combination of historical processes.[29]

As discussed in Chapter 6, the first horse riders learned how to master huge tracts of land and travel at relatively high speeds, and all with the viewer's perspective of about ten feet above ground. The shift in consciousness must have been huge considering the newly found confidence, power and sense of space, and it has been claimed that the spread of ideas of Lao Tzu, the Buddha, and the Hebrew prophets were propagated because of horse-riding.[30] There is also the neurological union of horse and rider, discussed in Chapter 6 which might lend itself to this theory. To ride and control a horse is not a simple human experience but that of connection with an animal of immense strength and speed. The control of the vast Caucasian steppes and herds of domesticated (or semi-domesticated) animals as a conscious experience may or may not have contributed to the new spiritual shift; but almost certainly spread new ideas.

The Axial Age came about just like agriculture did thousands of years before. It was a polygenetic development, happening in many places simultaneously. This turning point in human cultural evolution occurred between 800 and 200 BCE. It concerns and includes the revolutionary thought of Confucius, Lao-Tse, Mo Zi, Chuang-tse, and Lieh-tsu in China;

the writers of the Vedānta (also called the Upanishads) and the Buddha Gautama in India; of Zarathustra in Persia; Elijah, Isaiah, and Jeremiah in Palestine; and a whole host of thinkers in Greece. And this is just a select few! Incredibly, many of these lived at roughly the same times and in societies with little to no knowledge of each other.[31]

In terms of world history, the Axial Age was the last great change in human thought. It may even be understood as a change in the collective consciousness of humanity itself. The Axial period was a new time when people began to reflect, question, and trust in reason. They brought about new systems of ethics in their societies, and the first philosophers were born. They were the first humans to think about thought itself, to be truly self-reflective.[32] They developed historical outlooks which spanned more than the narrow spectrum of one set of seasons and the first historians were born. Thoughts became objects of thinking itself, and this was connected to spirituality. In addition to this, many of these thinkers were fascinated by movement.

Embodied beings as we are, we're physically bound to motion. Just like the water of Heraclitus' river, we're in no position to change the fact. On the flip-side, we're *spiritually* bound to its opposite, rest. At the deepest levels of what lies within us, there is the want—the need, even—to enter into a great stillness. These urges are present in all humans. The new form of society was failing to provide some spiritual achievement in its urban stillness, and its utopias may have been frustrating as they were impossibilities by definition.

Abandoning the City

All of this arose alongside a rejection of the city, possibly prompted by the marginalization of certain groups—slaves, women, and foreigners.[33] They reacted against the class structure, materialism, and power hierarchies around which the cities operated. Notwithstanding this, it would seem that spirituality was the driving force underpinning this movement against the materialism embodied in city life. In a mirror of the first instance of settling down, at the cusp of the Agricultural Revolution, it was spirituality again which caused this change. These thinkers returned to a nomadic and peripatetic life.

The Axial thinkers were in pursuit of stasis. Many lived an ascetic existence, renouncing the investment in material goods, and the entanglement that came out of that. In that search for stasis, they brought movement back into play, taking to nomadism and homelessness. Like an ancient, austere *flâneur*, they relied on self-examining consciousness coupled with new spirituality and rejected the perpetual buzz and commerce of the city.

They were after a deep stillness, like the type we may imagine emanating from Jesus of Nazareth, or the Buddha. They were deeply invested in the workings of the physical world around them and how best to live within the world as nature presented it, driving them to examine the moving and the static. They saw it not simply as Heraclitus saw the river, but also how Cratylus saw it—devoid of any thingness.

All told, the epoch of the Axial Age was nothing short of a shift in all human consciousness. People became aware of something transcendental in their experience in a way that provided an existential meaning to them.

Chapter 12

Logos

Now the existence of motion is asserted by all who have anything to say about nature, because they all concern themselves with the construction of the world and study the question of becoming and perishing, which processes could not come about without the existence of motion.[1]

—Aristotle

Much of Western civilization today is built upon the thought of the ancient Greeks. Their intellectual revolution began around the sixth century BCE, when the early philosophers, historians, and scientists began their practices. Western rationality, laws, ethics, and democracies are all rooted in that society. It was a part of the Axial Age that had arisen, one with a strong emphasis on self-awareness. A whole new mentality, and one that wasn't individualistic, it has been described as a "consciousness of the Greeks as Greeks, even though there was no political unity among them."[2] A broad social awareness seemed to awaken which included many schools of thought attempting to understand the universe in which they found themselves.

Spiritually, the Greeks were not at the behest of the weather gods like many of the cultures that preceded them. Anyone today reading the great works of Homer will find that the gods are remarkably human-like. They lack the powers that monotheistic gods have, and are easily swayed by ordinary desires such as jealousy and lust. Just like normal people, they are vulnerable to persuasion and make poor decisions on a whim. The world of Homeric gods is closer to a soap opera when compared to the all-powerful, all-seeing entities that later came with monotheism. As Xenophanes famously stated, if horses were to draw their gods, they would draw horses. But he is also recorded as saying something which echoes the static urge, "One god, greatest among gods and men ... always he remains in the same place, moving not at all; nor is it fitting for him to go to different places at different times, but without toil he shakes all things by the thought of his mind."[3] Here we see an early association between three elements: a mysterious, transcendental spirituality; a total absence of motion; and the power of mind.

There is a closeness between spirituality and stasis in much of the thought of these thinkers. For them, to explore the nature of the world meant exploring the nature of motion. And to explore the inner and metaphysical worlds was to examine mind itself. A later writer, Seneca, who was greatly influenced by many of the thinkers who will be the characters of this chapter, shows how the tension between kinesis and stasis in the ancient world was at the forefront of mind:

> All [travel] has ever done is distract us for a little while, through the novelty of our surroundings, like children fascinated by something they haven't come across before. The instability, moreover, of a mind which is seriously unwell, is aggravated by it, the motion itself increasing the fitfulness and restlessness. This explains why people, after setting out for a place with the greatest of enthusiasm, are often more enthusiastic about getting away from it; like migrant birds, they fly on, away even quicker than they came.[4]

He shows the endless lack of satisfaction embedded in kinesis, how we go from one thing to another, shallowly investing in moving out of restlessness, rather than curiosity or exploration. Seneca is suggesting we distance ourselves from a state of mind that is lost in endless distractions that kinetic restlessness demonstrates. Instead of being like this, he says that "Nothing ... is a better proof of a well ordered mind than a man's ability to stop just where he is and pass some time in his own company."[5] This static ability leads towards a greater happiness, he claims, in a sort of abandonment of the kinetic urge.

Ionia

The Greeks were on a mission to figure out how everything worked, basing their investigations upon experience and reason. Their culture was one of unending curiosity and huge investment in knowledge. They examined every phenomenon that was familiar to them. And although it was such a long time ago, we would be quite wrong to presume that their understanding was naive. The strength of ancient Greek thought can be measured in that Anaxagoras predicted the existence of DNA in the sixth century BCE, Democritus the atom, and Empedocles evolution in the following century. Even the Big Bang was suggested by Anaximander. All this demonstrates not just the profundity of their observation and rationalism, but the continuity of its effect throughout the centuries to our view of the world today. And two of the central approaches of investigation examined the nature of mankind, and that of motion.

It was, as the Greeks would have said, a shift in thinking from *mythos* to *logos*. *Logos* is essentially our relationship with the objective and factual. It describes scientific and pragmatic approaches to the world and during these Axial generations it replaced *mythos* as a primary form of mental discourse. Although *mythos*—the imaginative, emotional part of consciousness—didn't altogether disappear.[6]

Many of these thinkers felt that there was a single element from which all things came. Each had their own idea of what that might be—our good friend Heraclitus claimed it was fire. But for the earlier philosophers from the Ionian trading city of Miletus, such as Thales and Anaximenes, it was water and air. Through kinesis, Anaximenes argues, air *is* divinity. It's alive and always moving, and all things come from it.[7] Because it's always in motion, it imparts motion to all things. Whatever their differences in thought, there was a common understanding among many of these thinkers that there was some sort of unifying substance which governed the cosmos.

Another philosopher from Miletus, Anaximander, proposed the existence of an unknowable fundamental nature of the universe. Known by the Greek term *apeiron*, meaning "boundless" or "infinite," it was something that was unmeasurable, unknowable, and even lay beyond divinity itself. It existed apart from the classical elements (fire, water, air, earth) and it was the force that gives rise to both the creation and destruction of the universe.[8]

Anaximander had a particular interest in space, and was possibly the producer of the first map of the world. This demonstrates a massive shift in intellectual viewpoint, putting man rather than anything else as an overseer. It also was a manifestation of the division of space, as if the agricultural canals of earlier civilizations were now making their impact on the human mind. Axial thought in Greece was deeply involved with conceptions of space and motion, and much of this played out in geometry, not unlike the Sumerian culture before.

From the nearby island of Samos, their contemporary, Pythagoras made great leaps in mathematical and geometrical understanding. Of course, he is credited with things that he may not have been the first to discover, such as the famous theorem that had already been discovered by the Sumerians in city of Ur, predating the great mathematician and philosopher by about 1,500 years—mentioned in Chapter 5. Nevertheless, these attributions point to the gravitas the man had. His followers came to reject city living and became nomadic mendicants, in a pattern of behavior that reflected the cults dedicated to the gods Bacchus and Orpheus.[9]

What is most important in their geometry, is that it is an unchanging, static world. As mentioned in the previous chapter, the Pythagorean

Theorem holds true regardless of our existence, or lack thereof. Here is a reality where things do not change, contrary to the truth of Heraclitus' river.[10] It is a perfect world, as unreal as a utopia. Ironically, geometry is driven by the purity which can come from motion, such as perfect circles (of ripples in water) and straight lines (a falling object). Only in motion is there a perfect shape, but it's in a static world where we can examine these lines and shapes theoretically. Geometry happens in an alternate, utopian space, one which is fueled by second-order thinking.

Parmenides and Zeno

The three Milesian philosophers took it for granted that flux was naturally a part of the cosmological order, of the way things just happen to be. But when the next generation of philosophy erupted in Elea, everything was to be upended. The Eleatic school of Greek thought was based much farther west than Ionia (the area including Miletus and Samos). Led by the influential Parmenides in southern Italy, they did not take kinesis for granted.

Parmenides was so influential in fact, that it took until the great Athenian giants of philosophy (fifth and fourth centuries BCE), Socrates, Plato, and Aristotle for him to be overshadowed. He viewed the cosmos as a singular, unchanging entity, like Xenophanes' unmoving "god." He had a vision of "Being" that included not just physical movement, but all change. Typical of second-order thought, he argued that the Being encompasses all things, meaning that it must not be able to move. From that, he deduced that no change can exist. There is a single continuum which is perfect in and of itself.

> Being had no beginning nor can it have an end because it is firmly established, unmoved and eternal. It did not exist once in the past nor will it exist, since it exists at the same time all together, one and continuous.[11]

Here, we see that Being exists outside of time. He claims that it has never existed nor will exist at a single point in time. Without this to condition Being, nothing can ever move. Whatever movement we do experience must be an illusion, according to Parmenides. There is no reality as we see it, and no plurality of things. As humans, we see only glimpses of Being, and the rest of the time we don't see reality as it truly is. He argues that the universe is not made of things such as planets, stars and individual items, but that we only see it like this. It mostly exists beyond our human senses, so we can only understand it through our mental reasoning. Most suspiciously, Parmenides never accounts for how he came upon this knowledge.

Although it is easy to disagree with this, it's interesting to see his point of view as having this spiritual, and unifying quality of oneness. His description is of the cosmos as a single whole entity, static, perfect. And if any of this sounds a little familiar, it's because we touched upon it in Chapter 8. Henri Bergson wrote insightfully about time and human experience. He called our own personal subjective experience of time "duration," and like the Milesians, Bergson held movement to be more characteristic of reality than time.

I'd venture to suggest that Parmenides' idea of *Being* and impossibility of motion is simply too difficult to believe. It goes against our intuition of what life is really like. We live in a world where motion is happening all around us all the time, and it feels *real*. We can only trust our senses. Zeno, the best-known protégé of the great Eleatic philosopher, took issue with the idea of the impossibility of motion. But because of the standing of his teacher, he presented his argument to agree with his, and doing so resulted in a wonderful paradox.

There are a few examples, but probably the most famous is that of Achilles and the tortoise. Suppose that the great warrior-athlete is to race against a tortoise. He kindly gives the poor animal a head start to make up for the outrageous advantage that he has over it. At some point, due to Achilles' greater speed, we expect him to overtake the tortoise. But first he must catch up to the point where the tortoise began. And in the time it takes Achilles to get this far, the tortoise has progressed a little, and so remains ahead. If we continue to apply this logic, the warrior always gets closer, but the animal is always a little ahead. Zeno argues that Achilles can never overtake it.

Another paradox is known as Zeno's Arrow. It describes an arrow in motion, flying through the air towards a tree. At any given moment, the arrow can be thought of as stationary, as if frozen in time. Zeno argues that if the path of the arrow in motion is made up of a series of still, frozen unmoving moments, then the motion of the arrow is an illusion. Now if we think again of Eadweard Muybridge, whom we met a few times already, and his famous series of still images of a horse in motion, we can see some resemblance. We know the motion of both Zeno's arrow and Muybridge's horse to be real, so there must be a fault in the premise of the argument, that is to say a fault in the idea that time is divisible.

What Zeno's paradoxes point toward is a problem with the approach. We cannot keep dividing time and space. There is no reality to the frozen moments of Achilles, the tortoise, or the arrow, just like there aren't frozen moments of a horse in mid-stride, with all its feet off the ground. Motion is a *phenomenon*, not an object, and because the act is dependent upon space, it is therefore confused with it being an object existing as part of space rather than passing through it. Bergson himself wrote about this at length concluding that indivisibility cannot arise.[12] Movement is only real to us when it is embodied, when we engage with it ourselves. We are better off trusting our senses when we see or feel movement in action, and that is precisely what Aristotle instructed us to do when he wrote about these arguments. If we cannot trust our senses, then anything is valid.

Muybridge's photos of the horse in motion are also very compelling. They appear to depict reality, and it's a reality that we can easily connect with precisely because we so often deny the truth of Heraclitus' river. But there is no such thing as a horse suspended in the air; we know this to be absurd. The images which comprise cinema have had their motion—and therefore their reality—taken from them by the camera, only to be returned by the wheel of the projector when played in sequence. If we *only* trust our senses, then also anything can be valid. It is similar when we are emotionally moved by the stories that make up the film. Zeno's paradoxes teach us that we must use a critical mind.

What much of this means, is that the combination of reason and experience wasn't enough. Most, if not all of these thinkers ended up proposing that gods exist, or reality was not as it seems to us, or that there is some divine nature to things—such as *apeiron*—things which we don't have a direct experience of. *Mythos* and *logos* together weren't enough. Something more was needed to understand the world they lived in. And so, the next Greek to discuss is the great Heraclitus, who argued fiercely against the fixed, unchanging universe of Parmenides.

Heraclitus

Born in approximately 535 BCE in Ephesus to a wealthy family of the ruling class, Heraclitus left the preordained path of political responsibilities, later saying that he had gone in search of himself.[13] The little we know about his life demonstrates that he lived out his philosophy in an ascetic way. This would become a key characteristic of many of the early Greek philosophers. It was an applicable vocation, not merely a theoretical one as we might witness it today. Like Socrates after him, Heraclitus acted as an instigator as well as an investigator, his mission being to "prod sleeping minds to the waking state."[14]

Heraclitus was perhaps the first known person in the West to express these ideas of motion the way he did, having it as central to the nature of reality. It is why I started the first chapter with his most famous fragment, that one cannot step into the same river twice.

Although there is little information about Parmenides' life, the two were probably contemporaries. But Heraclitus had a completely different vision of the world when compared to him. Where the great Eleatic claimed that "the ultimate truth is an unmoving and unchanging being," Heraclitus argued that the "ultimate truth of the cosmos [is] continuous change."[15] You may remember from Chapter 8 that this was echoed many centuries later when Bergson stated that "reality is mobility."[16]

It is hard for us to argue with Heraclitus here. When we feel that we are still, it is because we have forgotten the spinning of the Earth around its axis. We have forgotten the orbit of the Earth around the Sun, like all the bodies in the Solar System, and that of the Solar System around the center of our galaxy. Even Heraclitus' book—which unfortunately has not survived—had a title which translates best as *On the Moving World*.[17] More importantly, it is in Heraclitus that we find the union of the investigations of kinesis with a more developed idea of the *logos*.

Reaching further than *Being*, the *Logos* of Heraclitus is the directing agent of the eternal flux throughout the universe (for clarity I shall capitalize Heraclitus' *Logos*, as distinct from *logos*). Nothing has the capacity to remain fixed. *Logos* is the only exception to this law. It *is* unchangeability. We could maybe say that it is the cosmos itself. Because of the constant flux of life itself, Cratylus brought it further, suggesting that the river itself doesn't exist. There cannot be any fixed thing, if all things are naturally and essentially subject to change. On the surface, it may appear otherwise, and it is here where our illusions lie. Zeno showed that time isn't divisible. Now, if he's correct, it means that motion isn't divisible either. Heraclitus deducts that nothing *can* be still. No fixity, no stasis, no thingness.

Heraclitus suggests that the river does exist, at the very least by mentioning it in the first place. But then Cratylus suggests that it doesn't. In different ways each philosopher is difficult to argue with. Thus, the Indian philosopher and yogi, Sri Aurobindo explains that "the waters into which we step, are and are not the same; our own existence is an eternity and an inconsistent transience; we are and we are not."[18] In short, *we can and cannot step into the same river.*

Here we are introduced to something far more mysterious indeed. The mental twist in getting one's head around what Aurobindo suggests demonstrates exactly what Heraclitus' *Logos* is all about. It's only partly rational. We must hold two opposing ideas at once. But Heraclitus is here to help. Even though so little survives of his writing, his aphoristic style points us towards opening up to the *Logos*. Another fragment tells us that "the way up and the way down is one and the same."[19] Here we are shown the unity of opposites in a real-life metaphor.

Like Parmenides' *Being*, his *Logos* is also unconditioned by time, and therefore unconditioned by motion. It exists as a unity, a singular integral whole. Our rationality by comparison is a mediocre tool of thought at best. *Logos* means a stillness derived from balanced tension. Its unmovingness balances against the flux and buzz of life. In a way we could understand *Logos* as a combination of three things: second-order thinking; the *apeiron* of Anaximander; and the *Being* of Parmenides.

Heraclitus offers a way to open up to the nature of *Logos*: "If we do not expect the unexpected we will not discover it, since it is not to be searched out and is difficult to apprehend."[20] In Fragment 34, he encourages us to use our own mental experiences in order to see the reality that he describes: "To be wise is one thing: to know that thought directs all things through all things."[21] We do this through something he calls *telos*, a natural spiritual reaction of turning towards our ever-changing mental experience. *Telos* can be thought of as the awareness of *Logos* in human consciousness. Both *Logos* and *telos* exist outside of the realm of constant flux. They lie somewhere a little beyond the world where his river is. We are drawn towards *telos* by how we must accept what we cannot deny, like aging and illness. Furthermore, it corresponds to our deep urge towards stasis, to transcend the kinesis that pervades the cosmos.

At the time, his reasoning was completely new and radically non–Western. In fact, it is strikingly familiar to many of the eastern spiritual traditions as we will discover in the following chapters. His predecessors had brought up the idea of a single "oneness," an indivisible wholeness to the universe that exists beyond the ordinary senses of the human being. But in Heraclitus, we have an always-changing universe. For our story of kinesis and stasis, the two are the same thing, he tells us in Fragment 52: "It rests by changing."[22] But how do we engage the *telos* so we can experience the reality of *Logos*?

The Athenians

It is all well and good to have a rational view of the physical world, but it was Socrates who applied this in a more Heraclitean way. He was interested in many things, but one which was highly important was stillness. Indeed, he made a name for himself during his younger days as a soldier by standing still for an entire day. This episode is repeated in Plato's *Symposium*, where the famous thinker was seen to be standing in the study since dawn.[23] He brought about a new attitude towards materialism, asking "Are you not ashamed that you give your attention to acquiring as much money as possible, and similarly with reputation and honour, and give no attention or thought to the truth and understanding and the perfection of your soul?"[24] Through his arguments, he unified the idea of stillness with that of renunciation, rejecting materialism, as Heraclitus had done before him. A combination of humility and search for wisdom was what would expose a person to *Logos*.

Socrates was not alone. A whole movement came after him, led by Plato and Aristotle. Other developments included the Cynics, who were led by a thinker heavily influenced by Socrates, Diogenes of Sinope. This man famously lived in a barrel on the streets of Athens, practically applying the new asceticism to his day-to-day life. But apart from the strangeness of their lifestyles, the Cynics lived by a new set of four simple values: honest speech, self-sufficiency, self-control, and humility.[25] Sitting in his barrel, Diogenes attempted to apply the values implied by a simple stillness to his life, rejecting the thingness that materialism and ownership lures us into believing. He moved little, and invested himself in gaining mental clarity. Regardless of what we may think about this today, he genuinely dedicated his entire life to this, once famously rejecting all kinds of riches offered him by Alexander the Great. Diogenes is said to have simply asked the great conqueror to step aside as he'd been blocking the sunlight. It is known that Alexander, so impressed by this, said that if he were not Alexander, he would like to be Diogenes.[26]

Having rejected materialism in favor of ideas, Diogenes didn't precisely revert to nomadism, but his life did somewhat resemble it with regards to his stance against materialism. Many others followed this way of life and it became a factor of inspiration for the Pythagoreans, Epicureans, and Peripatetics. All were a growth of this broader historical disillusionment with the city.

To settle the earlier disagreements as to motion being part of reality or not, it fell to Aristotle to make the point that there either must be constant motion, or constant stillness. It was he who gathered most of the propositions and theories of motion from the period of the ancient Greeks as a whole. It was all compiled in his *Physics*, where he discusses the merits and flaws of his predecessors' ideas on kinesis. Either Parmenides was correct, or Heraclitus was, but not both. Aristotle argues that there could be no middle ground between the two. He opens the final part of his *Physics* asking,

> Was there ever a becoming of motion before which it had no being, and is it perishing again so as to leave nothing in motion? Or are we to say that it never had any becoming and is not perishing, but always was and always will be? Is it in fact an immortal never-failing property of things that are, a sort of life as it were to all naturally consisted things?[27]

Here he argues that there can never have been a time where things were not in motion, because something must have been in motion to set off the kinesis in the first place. This means that motion must always have been and always will be. He takes this further, arguing that time itself is implicated here: "how can there be any time without the existence of motion?

...[It] follows that, if there is always time, motion must be eternal."[28] And later he surmises that "there never was a time where there was not motion, and never will be a time when there will not be motion."[29]

And motion also exists beyond time—given that there has never been a time without motion. Just like the other thinkers before him, Aristotle formulated a god-like omnipotent, omniscient, timeless power, whose existence we can perceive through the observation of movement. Things could never have all been still, because that would require a moving body to set everything in motion, which is a self-contradictory and absurd scenario. Logically, things must always have been moving. This puts motion itself as a *primary characteristic of the universe.*

Throughout the centuries, many thousands of individuals have found solace in the world-view of Heraclitus, and the values of Western civilization are firmly based upon the writings of Plato and Aristotle. The teaching of the cosmos being one of constant flux is undeniable, a self-evident truth. But it is through Heraclitus' *Logos* where part of this is taken further, to a spiritual understanding—one beyond mere rationality. The reality of *Logos* lies over and above ordinary human understanding, and there is one more Greek thinker who took it all a little further again.

Epicurus

Epicurus was born on Samos, the same island as Pythagoras, in 341 BCE. Like Seneca, he was attracted to the "well ordered mind," and thought that we ought to devote our lives to the simple pleasure of feeling well. This has often been mistaken to be a lust for reckless hedonistic pleasure, but Epicurus was clear about moderation. Genuine pleasure is found in *ataraxia*, a state of freedom from agitation, or strong desires; in short, a state of stasis where we are not itching to move somewhere else.

He is recorded to have used the words "kinetic" and "static" to describe two different forms of pleasure. Static pleasure is the absence of turmoil and pain, he tells us as he criticizes those who invest only in kinetic pleasures.[30] When considering suffering, the mental sort is far worse, according to this philosopher, so therefore greater pleasure comes in the static form, rather than the kinetic.

As a follower of other thinkers who were the first to propose the existence of atoms, he was very involved in studying the nature of matter and motion, and his understanding was uncanny in its resemblance to ours today. He also saw motion as a primary characteristic of the world, over time.[31] According to the later Roman poet, Lucretius, to whom we owe much of our knowledge of Epicurus, "Nothing from nothing ever yet was born."[32] This is a conclusion from an ever-changing state of matter, always in motion of some sort. Epicurus even took things to the level of Cratylus' river, claiming that "no solid form is found."[33] Nothing is really existing in a concrete sense. Our experience of stasis, then, must be one where we simply perceive kinesis. We attempt to be still, but aware of the changing nature of all things, at all times, and that is the greatest pleasure, the attainment of which is the purpose of life.

His keen-eyed observation of the world around him, coupled with reason, enabled him to see the law of entropy or tendency towards chaos, "each thing is quicker marred than made,"[34] again an encouragement to seek the pleasant in life. He figured out that at an atomic level, one that we could not see with the naked eye, things were at some point indivisible, and unpredictable.

Unlike many of his religious contemporaries, he intuited that there is no center to the universe.[35] From the earliest times, the atomist philosophers such as Democritus and Leucippus that inspired Epicurus developed their physical ideas in tandem with *ataraxia*.[36] It may be counterintuitive for us today, but there was a deep correlation between understanding how the physical world operates with the ethics of how to live a good life.

One of Epicurus' most innovative ideas was that atoms fell under their own weight. The universe is a material entity, he thought, but there had to exist some unpredictability, because if there wasn't, the world would be deterministic, with events already set out and we would be left with no free will. He supposed a type of tiny motion of atoms as they fell downwards, a little wobble or swerve. Lucretius later came up with the term *clinamen* for Epicurus' swerve. It brings our free will into the equation, resulting in an atomic dance, upon which the whole universe is built.[37]

Lucretius' extraordinary poem, *De Rerum Natura*, written in Rome more than two hundred years after Epicurus' death, even invokes the power of Epicurus' walking and wandering.[38] There is some directionless movement within the stasis, it seems, in a pensive and meditative way, not unlike Debord's urban drifting. It is an ethical development from the Heraclitean *Logos* uniting kinesis and stasis.

In the opening quote to this chapter, Aristotle points towards "the question of becoming and perishing." For the human being, conscious, sentient, who feels their agency and is free to move or examine the processes of the mind, both becoming and perishing are the forms which make up our world. This duality is reflected in our embodied structure of kinesis and stasis. But much of these ideas came to the Greeks—beginning in the trading town of Miletus—from much farther East where they originated and fed into the Ionians' thought. This is where the deep desire to transcend the duality of kinesis and stasis arose.

CHAPTER 13

Tao

The stillness in stillness is not the real stillness. Only when there is stillness in movement can the spiritual rhythm appear which pervades heaven and earth.[1]

—Anonymous

The Greeks were not the only culture collectively striving towards a transcendence of kinesis. In China, a new set of traditions simultaneously took hold, but with more mysterious results. There was a similar cultural push towards an experience of stasis, beyond the regular mundane life. Not unlike the advent of agriculture thousands of years before, it appears to have again been a coincidence that these things happened concurrently with other civilizations.

Just like Ionian Greece, it was boom time in China during the late Zhou Dynasty. Since the eighth century BCE, new city-states had been forming in central China, and trade flourished. This was one of the most important conditions for the flowering of the Axial Age, as with trade came a new tradition of wandering spiritual teachers.

Scholars typically divide the Zhou Dynasty into two periods. The Western Zhou lasted from the mid-eleventh century BCE up to the moving of their capital to Wangcheng, a city farther to the East in 711 BCE. It was here, during the Eastern Zhou, where the Chinese Axial Age took hold. The Eastern Zhou is further divided into the Spring and Autumn Period (770–476 BCE)—named after an early chronicle supposedly composed by Confucius, and the period of the chariot graves in Chapter 4—and the Warring States Period (476–221 BCE), when the great master actually lived.

The early Zhou rulers had introduced a set of rites as a cultural tradition, which could be performed before the spirits. And it was during the blossoming economy of these city-states that a new spiritual development began to take place. The lineage of monarchs who started their dynasty introduced what was then a new concept, a "mandate of heaven." This was a forged agreement between the ruling class and the gods. It was a simple way to unify the people and enforce, from higher authority, a system of social rules by which they could easily live. What they developed was the belief in a spectrum between ordinary people and the gods.[2] Rather than a gulf existing between the heavens and the Earth, there was a continuum between the two. Humans, as ethically responsible and intelligent beings, were cut from the same cloth as the higher powers, not unlike the gods as presented by Homer. The Zhou ruler, whoever they may be, had a divine power conferred on them directly from heaven. But ordinary people were also able to perform sacred acts, through the central act of Zhou religious life which was based upon ritual.

This, the longest-lasting of Chinese dynasties, was a fountain of culture with a deep richness in philosophy. It was widespread throughout such a vast area and contained so many various and conflicting ideas that it became known as the Hundred Schools of Thought. Just like in Greece, it was an era of itinerant scholars. They left the settled life, and moved in thought equal to their nomadism. Through their peripatetic wanderings their ideas went far and wide, and this is demonstrable from the many cross-influences of each school on another. All told, it was a geographical web of thought, ideas, and new theories all taken about on foot.

Sadly, the era ended in 213–2 BCE with a purge against intellectualism, most symbolically demonstrated by an infamous event known as "the burning of books and burying of scholars."[3] With this, the Zhou Dynasty ended and the Qin Dynasty began, uniting all of China under a single emperor for the first time. Although the effort was made to destroy the knowledge of the Hundred Schools, plenty does survive.

Kongzi

The greatest of these itinerant thinkers was Confucius (551–479 BCE) also known as Kongzi. A great spiritual and social philosopher, he lived the peripatetic life, giving advice to many senior civil figures and formulating an ideal society. His utopian vision, like so many others, looked back like Plato's had done, to the values of the Western Zhou, particularly their ritualistic spirituality, which he saw as flawless.

His worldview mirrored that of many early Greeks, interpreting the world as a non-created and godless place which lacked an overarching dominant power. The broader cosmos, as Confucius saw it, was naturally self-regulating and maintained a balance between all its various elements. It worked like a giant mechanical clock, operating from the motion guided by the self-evident laws of nature.

Although he is normally thought of as a social philosopher who stressed the importance of education, Confucius was also a deeply spiritual man.[4]

He carefully formulated a path for people to follow, insisting that it was our responsibility to do so. Our actions were to be the driving force for development, not our inherent nature. According to him, it is up to us humans to hold to an ethical practice. As we develop, we may become a *junzi* (superior man), and beyond that a *shengren* (sage). Further again, one may attain the highest level, known as *ren*. Most scholars today keep to the original word, *ren*, as it is considered untranslatable. More than simply a set of personality traits to adopt, it is marked by the typical behavior of the individual in question. *Ren* can be thought of as behavior itself, rather than the person behind it. Thus, the scholar Mo Zi remarks upon their rarity, "There are many teachers in the world, but only a few who are *ren*."[5]

It is because of such a system of ethical development that he envisaged education as a central element for society. Confucius was fiercely against learning by rote. Education is important because it is how we learn our morality, he argued. Through this, one may become a *junzi*. But without a desire to learn, moral perfection will not come our way.

This and many more of his ideas were brought into fruition by his best-known follower, Mencius (371–289 BCE). At the center of his interpretation was the idea that human nature was inherently wise and altruistic, albeit not always on display. "No man is devoid of a heart-mind sensitive to the suffering of others," he tells us.[6] We are equally born with a deep spiritual potential and it is our responsibility to nurture this.

Mencius probably took this idea from an earlier text, the *Analects*, one of the chief texts of Confucianism, although it is not totally clear if it was really him who composed it. In it, there can be found a utopian system of socio-political order, the human realm being the only part of the universe which needs some sort of order, lest it fall to savagery. This need is a consequence of the style of city-state living the world over, as can be seen by the prevalence of Confucius' central answer to the problem. "What you do not want others to do to you, do not do to others," he says in the *Analects*.[7] This has since become known as the "Golden Rule," and is found in many religious traditions.[8]

This simple rule was central to all of the Axial Age movements, and is distinctly second-order, in that one must imagine oneself from outside. The exercising of empathy in this intellectual sense would not have been necessary in prehistoric life. And perhaps it is not a coincidence that it took urban nomads to come up with it. To us today, it's a no-brainer, such was their influence. It even forms the central pragmatic element to Christianity at its most basic level, and in many ways is a common method of making a difficult ethical decision for many Christians and non-Christians. It holds as a self-evident moral rule. Probably at the time, though, this was rather groundbreaking, for it had not always been the case.

The Confucian way is about interaction with others, learning from and benefiting them, always encouraging others' self-cultivation. In the inner aspect it's about attainment of wisdom and peace—becoming a "sage within." It is to become still, without wants or needs, and to attain a lack of restlessness—what Epicurus called *ataraxia*. The Confucian understanding of socio-political order was fused with a stillness and spirituality. This deeply kept association with the transcendence of movement was to inform a later successful development.

What Confucius envisioned became known as the *Tao* ("way"), and it is no accident that this same word would later become used in one of the most important spiritual systems of the Axial Age. Many aspects of the Taoist tradition which followed the Confucians are based on the early teachings of Kongzi himself, as well as his principal followers, Mencius and Xunzi.

Over the centuries that followed, deep changes came through Confucianism. Like the shift from early Greek *logos* to the Heraclitean *Logos*, the "way" of Confucius underwent a change towards the unknowable, the still, the unmoving. But this wasn't all appropriated onto the master's teachings. We do find a seed of it in the *Analects*, where he gives a hint at the wisdom of ruling by non-action. "The Master said: 'Shun was certainly one of these who knew how to govern by inactivity. How did he do it? He sat reverently on the throne, facing south—and that was all.'"[9]

Wu-Wei

Part of the Confucian spirit of social order is the idea of a king or government ruling through the practice of non-action (*wu-wei*), and it was to become a central idea of the spiritual movement of Taoism which followed. Stillness in action, through the practice of *wu-wei*, played a central role in both of the core texts of Taoism, the *Chuang Tzu* and the better-known *Tao Te Ching*.

A later school, the wonderfully named Way of Mysterious Learning, held that *wu* ("nothingness") was central to the nature of ultimate reality. Understanding this means doing away with words altogether. Truth is a void, according to Wang Bi, one of this school's most prominent philosophers. "*Wu* has neither form, nor shadow; it conforms completely to what surrounds it.... Its form is invisible: it is the Supreme Being."[10] He stressed that *wu-wei* can lead to laziness and aimlessness, and we need to be wary of this. His belief was that despite the invisibility of what he calls a supreme being, and the difficulty in understanding it, in attempting to know it we can learn how to live better. This matched with the ethics of

Confucius. *Wu-wei* is the naturally ethical act which comes of this. Wang Bi believed that Confucius was the embodiment of this quality himself as he believed him to have been a true sage. And so, the teachings of Confucius and his followers were eventually adopted into the Taoist system which succeeded it. Meaning in this context was something that would prove to be far more elusive indeed, in that it inherently resists definition, putting us in a realm with no straight answers.

The concept of *wu-wei* held, though, and lent some importance to meditation. The practice of literally not moving, being physically still, and not reacting to the events that occur in one's consciousness was one way to implement *wu-wei*. It means kinesis in mind only, and the less one moves physically the more one develops statically. The more he or she opens up to the mysterious *Tao*.

The eponymous *Chuang Tzu* gives us a colorful example of this traveling without moving, "Which of us can be where there is no being with, be for where there is no being for? Which of us are able to climb the sky and roam the mists and go whirling into the infinite, living forgetful of each other for ever and ever?"[11] This passage implies that the realization of a higher spiritual existence is purely movement, without substance. Just like the river of Heraclitus, there is only a flux, here described as a "whirling into the infinite." The performance of impossible physical feats, such as climbing the sky, suggests a non-physical motion, leading to realization of the *Tao*. And again, like the early Greeks intuited, at some level there is a oneness to all things, a lack of tension between the transcendental and the mundane, stasis and kinesis.

Tao operates like Heraclitus' *Logos*. It is elusive by its very nature, resisting any attempt to catch it, or describe it. It is mysteriousness itself. As the *Tao Te Ching* states: "The Tao seems nonexistent, but it is the basis of existence."[12] It is beyond understanding. This important text of Taoism, attributed to Lao Tzu, opens with the following statement:

> The Tao that can be understood
> is not the eternal, cosmic Tao,
> just as an idea that can be expressed in words
> is not the infinite idea.[13]

This is one of the central themes of Taoism, the idea that the ultimate characteristics of nature lie outside rational human understanding. It mirrors very closely Socrates' notion that wisdom takes the form of not knowing ("I do not think that I know what I do not know."[14]) We are stuck using mundane, worldly language to communicate it as we can do no better. But our words aren't the things that they signify. Words are just words, and so the *Tao* that we describe cannot be the real *Tao*.

It was really the later generations who took Confucius' idea of a systemic social order, and developed it spiritually. The *Tao* of Taoism is different to the *Tao* of Confucius. It is an unmoving, serene and static reality that lies far beyond our comprehension. Like the early *logos*, it wasn't enough just as it was. Because rationalism, good as it is, fails us when we go into the more esoteric emotional and spiritual aspects of ourselves, something else is needed.

Regarding our proverbial river, the Chinese Axial culture was approaching kinesis in a manner far closer to Cratylus than Heraclitus, to put it simply. There is far more to it than waters ever flowing and a river ever changing, according to the Taoist texts. The river may not exist at all. Taoist stasis gets us closer to experiencing the immateriality of things. Xunzi therefore amplified the importance of stillness as the goal of the spiritual life:

> How does man know the Way? By the heart. How does the heart know? By being empty, unified and still. The heart never ceases to store, yet something in it is to be called empty; to be multiple, yet something in it is to be called unified; to move, yet something in it is to be called still.... The heart when sleeping dreams, when idling takes its own course, when employed makes plans so never ceases to move, yet something in it is to be called still—not letting dream and play disorder knowledge is called being still.[15]

This practice of *wu-wei* seems to be anathema to kinesis. We are born movers, designed perfectly for walking and running through countless millennia of mutations and natural selection. But the Taoists (and even some others such as Socrates) had this deep interest in sitting still. Doing it was part of our nature, they claimed. The point is to transcend kinesis, to overcome it in an attempt of higher experiences of spirituality through consciousness. So, the motion becomes transferred from the physical to the mental.

In the Confucian model, we need to work to develop into a *junzi*, *shengren* or a *ren*. It is part of our nature to turn towards these patterns of behavior, we are drawn in that direction naturally. But that does not mean that we can attain it with no effort. That is achieved through non-action, paradoxically. It is in this that the two deep desires described many times throughout this book are in most tender conflict—to move and to be still. The early Taoists believed that it may be achieved through *wu-wei*.

Another example is that of *qi* (pronounced "chee"). There is no simple definition of *qi* but Mencius refers to it as a "life force" which pervades a person, which has another higher, more refined dimension to its essence.[16] Confucians believed this vital energy to be neither material nor immaterial.[17] It is something else, part of the cosmic system of natural balance that we have yet to understand. It's the flow of *qi* rather than the *qi* itself which substantiates it, its flow making it the "fountain of life."[18]

Guan

Another classical text, the *Guanzi*, contains a discourse on meditation techniques known as the *Neiye*. They are primarily focused on sitting still and placing one's attention upon the breath, all the time attending to the constant motion caused by it.[19] The *Guanzi* describes sitting posture, concentration on *qi* and the ultimate spiritual fulfillment if practiced enough, making one a sage. Other texts such as the *Chuang Tzu* also relate this. One develops an awareness of *qi*, through doing this practice, and with many years of experience may finally enter an alternate state of consciousness, known as *guan*.[20]

This word, *guan*, refers to a state of lucidity in the observation of self. It is the experience of seeing oneself within one's context directly rather than through ideas of how things ought to be. It is an impartial view which empowers the meditator especially in difficult situations, because he or she can see things more clearly than before, and behave accordingly. One of the elements of this is a deeper awareness of the constant state of change, or Heraclitean flux. The result of all this is a deep knowledge of a world of movement instead of fixed entities and ideas, which allows us to go further into our inner being and understand the nature of the universe.[21]

Many Westerners today may feel that the practices of non-action and meditation may simply cause boredom. We live in a world of constant distraction which turns us away from these experiences at any cost, caused by the denial of the flux of all things. But this is a very recent phenomenon, and the fear of boredom dates back only to the seventeenth century.[22] This, of course, is the era where individualism became central in those societies, and has only gotten stronger as a cultural force. We feel the need to be constantly *doing* something, or being entertained. Time is money, as they say, and our existence ought to be justified somehow by action. This couldn't be any further from the Taoist way.

But there is an opening up of sorts that arises out of the state of boredom. If we can sit it out, we quickly bring an interest to our direct experience. As one German writer claims, "A landscape appears in which colorful peacocks strut about and images of people suffused with soul come into view... [Your soul is] swelling, and in ecstasy you name what you have always lacked: the great passion. Were this passion—which shimmers like a comet—to descend, were it to envelop you, the others, and the world—oh, then boredom would come to an end."[23] Here we find boredom leading to its very opposite. It is in this type of experience that the human mind opens up to a glimpse of the *Tao*. In sitting still, we observe how active the mind can be.

Ironically, we must use our bodies to not move. The sensations of life come to our minds through our bodies. Despite their itch to move, and their being undeniably perfect for moving, it's advantageous to be still, say the Taoists. Our experience of the world is the way it is precisely because we are embodied. The body, however, means that our self is limited, and only when we are free of it, shall we be able to really meet the true reality. The *Tao Te Ching* tells us that "The reason I have great trouble is that I have a body. When I no longer have a body what trouble have I?"[24] Only after being freed from our bodies are we free from kinesis. We can occupy whatever shape or place we want, free of any restriction in time or space. Through death we overcome kinesis, our embodiment, and even our entanglement with it. But for the time being, here on Earth while we enjoy our stay, the closest we shall get to experiencing the *Tao* will have to be through our embodiment. Thus, we find a famous aphorism about a certain monk, Yün-men. When was asked by a junior monk, "What is the Tao?" he answered, "Walk on!"[25]

Within the stillness of *guan* is the possibility to glimpse something beyond the mundane. Through the new consciousness that arose at the dawn of the Axial Age, these philosophers, rejecting the materialism of their societies, were trying to go further in mind. They were transcending kinesis, or at least attempting to do so, not merely acknowledging the desire for a final resting point. Longing for stasis, the drive was to be free of the itch of movement.

In one manifestation of *guan* there is eternal motion, and in the other there is the ultimate reality, a *Logos* or *Tao*, unmoving, perfect, and wholly integral. Humans can respond to this fact in only two ways, by addressing their urge to move, or their urge to be still. In the Chinese tradition, it all starts with the eternal motion throughout the universe. And the seed of it all is in the *I Ching*, also known as the *Book of Changes*. This predates Confucius, and after him it was included as one of the "Five Classics" which made up the canon of Confucian literature.

This ancient text was likely the only one to survive the "burning of the books" purge.[26] It is often mistakenly seen as a method of divination, but is more likely to have been used as a tool to aid making decisions.[27] It is based on the idea of constant motion, change affecting change affecting change. Its fundamental concept is to open us up to the reality of unceasing kinesis, focusing us not on things in a temporal state of being, but on their continual motion, disconnecting us from the thingness we see in the world around us.[28] One of the reasons it has survived until today is because of its importance to the Way of Mysterious Learning school.

The flux of *qi* is present in the constant motion of things. According to one prominent scholar, the focus of the *I Ching* is "not on things in their state of being—as is chiefly the case in the Occident—but upon their movements in change."[29] The text itself describes one type of change as "nonchange," a sort of static changelessness, but one defined as change nevertheless, a seeming paradox. "Nonchange is the background ... against which change is possible," we are told.[30] There has to be a fixed point, according to this logic, from which movement is made, as if it's *dependent* on there being an observer.

The Way

From the perspective of ultimate reality, the nature of kinesis is something that lives in the consciousness. Movement is an incomprehensible illusion that we suffer. The quote at the beginning of this chapter is taken from a medieval text and later was popularized by no less than Bruce Lee. In it we find a similar explanation. It is stillness in movement, rather than stillness in stillness which is attainable, a neutralizing unity of opposites. Elsewhere, in the *Tao Te Ching*, the reality of the *Tao* is unmoving, and ultimately incomprehensible and indescribable:

> [The Tao] is unseen because it is colorless;
> it is unheard because it is silent;
> if you try to grasp it, it will elude you,
> because it has no form.[31]

In order to realize a true stillness, we would need to balance the relevant forces. We find this precisely echoed in Newton's Third Law of motion which states that for every action there exists an equal and opposite reaction. In Taoism we find its philosophical and spiritual counterpart.

The early Taoists believed that any exercising of power would bring about an opposite force, spiraling the motion, and taking one further away from the state of stillness. In an echo of Heraclitus—who claimed that "the way up and the way down is one and the same"—the Taoists saw reality as various unions of opposites.[32] This is depicted in the well-known symbol of Yin and Yang. Seemingly, both are needed to exist so that the universe thrives. It is the holistic and ubiquitous existence of both that creates the oneness of all things. In the words of the historian Arnold Toynbee,

> This alternating rhythm of static and dynamic, of movement and pause and movement, has been regarded by many observers in many different ages as something fundamental in the nature of the Universe. In their pregnant imagery the sages of the Sinic [Chinese] Society described these alternations in terms of Yin and Yang—Yin the static and Yang the dynamic.[33]

The Yin-Yang symbol relates most obviously to a balance and interrelation between opposites. But it also is a manifestation of Confucius' Golden

Rule of treating others as you wish to be treated. It graphically shows the interconnectedness of actions and their consequences. And it works as a reminder for ethical behavior. This is the overarching ethical teaching of the Axial Age, based on an understanding of the laws of nature—rediscovered later through physics by Newton in his Third Law. It is found in the teachings of Jesus of Nazareth, the Buddha, and other philosophers of the Hundred Schools, such as Mo Zi. An empathy-based altruism is an incredibly efficient rule to allow for an easily operating society. We simply must act towards others in a manner which we would like them to act towards us. Today's parlance has it morphed somewhat into the phrase, "what goes around comes around."

Fig. 15. Image of Yin and Yang, complete with the Eight Trigrams derived from a broken line representing Yin, and the unbroken line Yang. In Taoist cosmology, they represent the elements, clockwise from top: sky, wind, water, mountain, earth, thunder, fire, lake.

The reconciling of opposing entities is what we may do to begin to understand the nature of *Tao*, and this applies to the kinetic-static duality too. Hence we have plenty of examples through the Taoist texts showing the reality of both eternal kinesis and the simultaneous absence of it. It also reflects the duality I proposed in Chapter 1 between Heraclitus and Cratylus.

The role of the observer of motion is important here, just as much as that of the agent of the Golden Rule. Indeed, if movement is genuinely fundamental to reality, it doesn't exist outside of the objects that move, instead it's intrinsic to them.[34] This is a major theme of twentieth-century quantum physics, that observation plays a part in how things work, but that is something that was known through spiritual tradition so long ago in Chinese culture. Thus, "The ancient Chinese mind contemplates the cosmos in a way comparable to that of the modern physicist" says none other than Carl Jung. Modern physicists, he continues, "cannot deny that [their] model of the world is a decidedly psychophysical structure."[35] But rather than seek out a fixed, objective model of kinetic or static reality, we are better off if the answers that science provides are not definitive. One modern physicist considers them more reliable as such,[36] and another, one of the greatest of the last century, Niels Bohr, included the Yin-Yang symbol when designing his coat of arms when he was awarded the Order of the Elephant, the highest honor of his native Denmark. Accompanying the symbol was the short text *contraria sunt complementa* ("opposites are complementary").[37]

We are thrown by the inability to intellectually reconcile this conundrum. And that is the point—our *logos* or the Confucian social order isn't enough. We cannot reconcile our two conflicting urges mentally. At some point we have nothing left to do but resign ourselves to it. This is why Chuang Tzu, after claiming that "The *Tao* never alters," later instructs us, "Let us forget the lapse of time; let us forget the conflict of opinions. Let us make our appeal to the Infinite, and take up our position there."[38]

The static envelopes the kinetic, with *Tao* meaning the "way," suggesting that kinesis must be part of our search for stasis. Perhaps it's not unlike walking an ancient pilgrimage route (like those in Chapter 2), or following the desire lines of Chapter 10. Perhaps it could be thought of in terms of the colonizing of the Earth, but brought into another realm.

But for the time being, here we are in the human world, where movement not only is the law of nature (as it is according to Heraclitus), but is the source of the very mystery of existence. If the *Tao* is unmoving, where does movement ultimately come from? asks Chuang Tzu. "The winds come from the north, going first to west then to east, swirling up on high, to go who knows where? Whose breath are they? Who, doing nothing, creates all this activity?"[39] Perhaps he is borrowing this idea from the *Tao Te Ching* which states, "The universe, the earth, and everything in it comes from existence, but existence comes from nonexistence."[40]

Regarding who's behind the blowing of the wind, there's no telling, as Taoism operates via the mystic rather than the rational. In a profound sense, this is true of everything. Existence may come from *wu-wei* itself, but we cannot know. Even the sense of self is fickle, because of the law of eternal change, according to Chuang Tzu. In a remarkably eloquent passage he tells us,

> Once upon a time, I, Chuang Tzu, dreamt that I was a butterfly, flitting around and enjoying myself. I had no idea I was Chuang Tzu. Then suddenly I woke up and was Chuang Tzu again. But I could not tell, had I been Chuang Tzu dreaming I was a butterfly, or a butterfly dreaming I was now Chuang Tzu? However, there must be some sort of difference between Chuang Tzu and a butterfly! We call this the transformation of things.[41]

We have lost what we thought was inherent to our nature, our very selves. Like other Axial Age traditions, there appears a utopia in ancient Chinese spiritual philosophy, and not only in the system of Confucius. It looks back, not forward, like the Garden of Eden of the Judeo-Christian tradition. Chuang Tzu himself speaks of it, describing life before the settlement of agricultural peoples, relating a time of abundance when private property did not exist and how easy that made it to practice *wu-wei*.[42] The nomadic life that long ago preceded the Axial Age city-states was looked upon as a utopia, just as it was seen by the ancient Greeks. One among the Hundred Schools of Thought based itself completely on this idea, purporting to agriculturalize the utopian past.[43] The *Tao Te Ching* has one chapter devoted to a utopian description, even allowing its inhabitants to "have carts and boats, but there is no reason to ride them."[44]

A new but similar worldview would grow to dominate China, that of Buddhism. It began to take hold in the area about the first century CE. Although it would take about another 400 years for the Mahayana interpretation of the Buddha's teachings to blossom, it would deeply diminish the impact of Confucianism and Taoism. But, in India it had already been spreading since the days of Confucius himself.

CHAPTER 14

Nirvana

Unchanging things don't change, and changing things do change—until they change into things that don't.[1]

—Addy Pross

Since ancient times, the study of motion has consumed and fascinated some of the greatest minds of humanity, such as Isaac Newton and Albert Einstein. The field of physics has produced countless theories, some of which may stand correct for centuries. For example, the equations of Newton—discovered in the mid-seventeenth century—were used by the European Space Agency team that carried out the Rosetta mission to land a spacecraft on a comet in 2014.[2]

We have seen a parallel between the study of motion and the emergence of a second-order consciousness at the time of the Axial Age. Rationality was the new currency of thought in these times, perhaps difficult to imagine as we are so conditioned by it in our day and age. Part of this way of thinking was invested in the study of motion. But the Axial thinkers also realized that rationality wasn't enough. It didn't solve the deeper questions of the universe so easily, those that relate to the emotional and spiritual aspects of humanity. A great mental tool that it was, it lacked the ability to access the other mystical dimension that we sense from time to time. But the study of motion in the scientific sense paved the way for the Axial Age mind to understand its reality.

Sometimes, the world we live in appears only to exist outside of us, and everything behaves in a predictable way. Our kinesis leads to other kinetic events. I kick a ball and off it goes. Other models of physics show different realities, so to speak, where things are softer, more pliable, and our being part of the kinetic world has an effect on how that world behaves. In Einstein's relativity, the experience one has of time and space depends on how fast they are moving. In quantum physics the motion of a particle may be influenced by the fact that it is being observed.

What this means is that the goings-on around us don't exist independently of our actions. Aristotle was correct when he wound back the clock and made the assumption that all motion comes from previous motion. And if we put ourselves into that equation, we see that our kinesis affects our surroundings. We have agency in our world, we're connected with it, and part of it.

Motion sometimes tells us otherwise. For example, when the railroads needed a standardized time, it gave people the notion of an independently existing time-world. Time ticked along by itself, it seemed, over and above ourselves. Before the nineteenth century, each town had its own time and the variations were tremendous. It was only due to safety and scheduling that a standardized time was put in place (see Chapter 8). Furthermore, public transport would arrive and leave (hopefully on time) regardless if an individual turned up or not. It gives the impression of an independent existence where people don't have agency. We give this more credence than it deserves because industrial motion is faceless and mechanistic. Trains and automobiles reinforce the thingness that we interpret in their existence. This is why Bergson's idea that we subjectively experience time as "duration" is important.

Entropy and Complexity

Time is not a man-made invention either. We've seen how Axial thinkers see it as a consequence of an ever-changing world. It also seems that time moves always in the same direction. We are getting older, day by day. If I knock a vase over and see it smash on the ground, I don't see it put itself back together. Time has a natural directionality (although at the quantum level it's not so simple). That direction is part of nature, the leaves falling from the trees demonstrating a greater cosmic law at play.

Scientifically, the law which explains this as a phenomenon is the second law of thermodynamics, or the law of entropy, which states that energy moves freely from order into disorder (the first law, mentioned in Chapter 5, states that energy cannot be created or destroyed). This was Ludwig Boltzmann's discovery in the 1870s. The falling vase breaks upon its impact with the ground, but we never witness it "unbreak" itself. It is not natural for the smithereens of the broken vase to come together by themselves. We have always known this intuitively, it's not that Boltzmann discovered some new phenomenon, but he did discover a way to express it.

The second law of thermodynamics states that energy, in the form of heat, always moves naturally from a hot place to a cold one. This will continue until a state of equilibrium is achieved. This is why a cup of tea left standing won't stay hot for more than a few minutes. Its energy streams out into the relatively colder air around it, until both achieve a balance at the same temperature. This state is called *maximum entropy*, and it is a cosmic law, meaning that it applies throughout the universe, just like Newton's law of gravity.

By nature, entropy is always increasing, according to the law. Heat disperses into a nearby cold area, and so the universe that we understand will eventually become a state of total entropy, where all energy will have scattered homogeneously. Theoretically, this ought to result in a state of eternal rest, or equilibrium, and would mean the "heat death" of the universe, as the average temperature across the entire cosmos gradually drops to something very close to absolute zero, and it becomes a static entity.[3] Interestingly, it took the existence of steam trains to bring about the discovery of entropy, because the more that was known about this area of physics, the more efficient the train engines could be.

But entropy is not the complete picture, as there seems to be an opposite force at play too. With enough time there is incredible growth of organisms and births of stars and galaxies. There is certainly a tendency towards increasing complexity, as atoms combine with the stickiness of carbon to form amino acids and simple proteins build themselves up to be complex organisms. Complexity kept increasing over countless millennia and immensely intricate animals such as reptiles and mammals came to dominate the Earth, with all their various organs and tissues and biochemical elements maintaining them. This shows a decrease in entropy meaning that structure increases, putting order to matter.

As Boltzmann himself is supposed to have asked, why was there so much order in the universe in the past?[4] If there is only entropy, how could there have been a vase, or a hot cup of tea in the first place? Down here on Earth, it is the faculty of self-replication through DNA which drives life. It makes sense that the gene's "desire" to replicate—so to speak—is the driving force behind the ordering of matter and the contradiction that we see to the law of entropy.[5] But none of this really means that we can escape the law of entropy entirely, just that there exist other forces simultaneously.

Both of these phenomena, replication and entropy, reinforce the idea that Cratylus suggests in saying that the river does not exist. The replication of DNA, its manner of mutating within that process, the coming together of certain types of matter, and the inevitable decay of all things mean that flux and thingness are both real, although they are contrary. The quality of thingness exists conditioned by time, and as the adage goes, "this too shall pass."

The British psychiatrist and literary scholar, Iain McGilchrist, argues that these two opposing forces are central to structure in cosmic terms:

> The most fundamental observation that one can make about the observable universe ... is that there are at all levels forces that tend to coherence and unification, and forces that tend to incoherence and separation. The tension between them seems to be an inalienable condition of existence, regardless of the level at which one contemplates it.[6]

It's as if this dualistic tension powers the universe itself. It powers motion, it powers our desires to attain stasis. What's remarkable is that we can see this dual nature from *within* the universe itself. That could be considered the great mental achievement of second-order thinking and of the Axial Age in general. It's precisely looking at things from this point of view, as if we were existing outside of time looking in, which is characteristic of Axial thought. We can't *really* take this position, of course, but if we hypothesize it, we can work out how many aspects of the universe work, which is exactly what physicists have been doing for much of this time, since the ancient thinkers. This stepping outside our world of operating within time is also what the Axial thinkers did in the spiritual sense too, bringing a quality of transcendence to the questions of kinesis and stasis.

Axial India

In India, during the Vedic times (approx. 1500–500 BCE), a concept of the universe emerged in which it is described as coming from a dynamic and primordial natural order. Despite what we may or may not experience, the underlying force governing the cosmos is one where the precarious balance is maintained by motion. This principle is known as *Ṛta* in Sanskrit (pronounced RI-tah), and belongs to the Vedas of Hinduism. In one of these texts, *Ṛta* is the "course of all things," it is pure movement itself.[7]

In later developments of Hinduism, one of the highest beings that exists is Shiva, who appears in the form of a "cosmic dancer." This supreme being is formless and transcendent, and dances eternally. His dance is both one of creation and destruction (mirroring our understanding of replication and entropy). The cycle of birth and death is a corresponding dance, and anything else merely a distraction, an illusion.[8] The only reality is flow. Interestingly, Shiva dances rather than moves. The reason for this is that he is the wholeness of the cosmos, leaving no place to move to. As one anthropologist explains, "I am everywhere and in everything: I am the sun and stars. I am time and space and I am He. When I am everywhere, where can I move? When there is no past and no future, and I am eternal existence, then where is time?"[9]

If the ultimate reality is like this, then all other phenomena and actions are going to tear us away from it. For this reason, many Hindus, Jains, and Buddhists have led a life of asceticism similar to that of Diogenes whom we met in Chapter 12. Their simple lives are based on the idea that a direct experience of the dancing cosmos is closer to them if they renounce worldly things. Many resort to nomadism, the city being rife with material temptations.

During the time of the Axial Age, the asceticism practiced in India was quite extreme. The Greeks were aware of these Eastern ascetics who wore rags and ate little food, the idea being to totally purify the body and mind from materiality and gain, as they were considered hindrances to our spiritual development.

One particular Greek visitor to India, Onesicritus, was a disciple of our Athenian barrel-dwelling friend, and a ship's pilot of Alexander the Great. Both he and Alexander met with one of their most-famed ascetics, Kalanos. Here the two men must have seen eye-to-eye in their worldview of rejecting the material life. But Kalanos even related the woes of leaving behind an itinerant utopian past:

> In the beginning, the world teemed with wheat and barley. Now it's mostly made of dirt. Once fountains yielded an abundance of water, milk—even honey, wine and olive oil. But excess and self-indulgence only made men insolent. In disgust at this state of affairs, Zeus took away these blessings and subjected man to a life of labor, when self-restraint and the other virtues developed, then opportunities for a good life reappeared. But greed and arrogance are once again threatening man's existence, and at present there is renewed risk of widespread devastation.[10]

So, here we find why these people committed to these practices, as a reaction to the indulgence which followed the large-scale food production and the inequality that came with it.[11] The long hours of sitting still in Indian forests was the exact opposite to this over-consumption. Recalling the utopian past of their ancestors, they viewed the entanglement with material things as undivorceable from greed. And since those early generations of agriculturalists, our environment has become steadily more manufactured and unnatural. The practice of asceticism was to disengage with what they believed was the cause of all this toil.

Stasis ties it all together. Firstly, there is the stasis of abandoning nomadism, the spiritual developments which brought about agriculture, which we are already familiar with from Chapter 5. Secondly, we have the utopianism that comes ultimately from the disillusionment with agriculture, the static cities and societies of the mind's invention. And thirdly, the meditative, ascetic lifestyle of these spiritual practitioners, following something not unlike the *wu-wei* of the Chinese thinkers.

Gautama

One of the most famous ascetics in India during the sixth century BCE was Gautama, the man who would later become the Buddha. He practiced an extreme form of self-imposed brutality, so much that legend says that if one pinched his stomach, they would feel his spine. This was some difference to the way he had developed as a youngster, as he was born into local royalty. Just like Heraclitus, he left that path of wealth and power behind him, irreversibly abandoning it to become a homeless wandering spiritual seeker thinking that it was a more meaningful way to live his life. On seeing that he, like everyone else, was vulnerable to old age, illness, and death, he realized that riches and power were just empty distractions from the realities of life.

After many years of this way of living, he finally made the realization that it was not helping in his quest for spiritual fulfillment, and so he abandoned his camp and fellow practitioners. He began taking more food again—to the disgust of the other ascetics—and again wandered the forests and villages of India, but still remained a homeless wanderer, going barefoot, begging for food, and practicing meditation.

According to tradition, Gautama gained enlightenment sitting under a tree after meditating deeply without interruption for many weeks.[12] What enlightenment means in the Buddhist tradition, is a mental state free from suffering. What's interesting about it, is that much of the Buddha's teaching has a mental emphasis. Most of the work cut out for the Buddhist must happen in their mind. We may even walk past an enlightened person on the street unwittingly.

At this moment of his enlightenment Gautama became the Buddha, the name meaning "the awakened one." After absorbing this experience for many weeks, he decided to return to instruct his former ascetic colleagues. Despite the fact that he had left them, he knew that they were relatively experienced practitioners and would be able to understand what he could relate of his experience. He traveled to Sarnath (which is still there today), and gave them the first of what would be many discourses that he gave throughout the rest of his long life.

The static phenomenon of existing outside of time, so essential to Axial thought, is illustrated in a traditional story from the life of the Buddha. In one account, dating back to the earliest writings of the Buddhist tradition, he is being followed by a bandit in a forest intent on murdering him. Once the chase begins, the sprinting bandit cannot keep up with the Buddha, who continues walking at his normal pace.[13] When ordered to stop, the Buddha complies, telling the murderous bandit that "I am standing still. It is you who are moving."[14] This is precisely the stasis sought after in Axial Age religions, one which allows some form of existence outside of the conditionality of time, and similar traits can be found in more recent religious traditions such as Islamic mysticism.[15]

Dharma

The central idea that lies at the core of the Buddha's teaching, or *Dharma*, is that all things arise and then pass away. Each thing or phenomenon that exists is conditioned to come into being, depending on external circumstances (themselves being conditioned in the same way), and then passes away with time. For example, sexual intercourse gives rise to infants, which at some point will die, due to the condition of their birth. We all inevitably die and having been born is a condition for this, as put wonderfully by Samuel Beckett in his one-liner, "Birth was the death of him."[16] We can say that the entire field of Buddhism—all the countless interpretations, schools, and traditions that have existed since that time and make up the Buddhist world—has germinated from this kernel idea. But the Buddha warned of overthinking this web of conditionality, saying instead that we should focus on what to do in this situation in order for us to gain enlightenment.

Like the rules of classical mechanics, conditionality is a law of existence rather than a theory. The more one investigates it, the truer it seems to be. Like Heraclitus' famous fragment we are sent thinking about that river whose waters are always new and ever-flowing. The more we question it, the more accurate it seems to become. The Buddha famously instructed his followers to test his teachings critically within their own experience—all that was needed was an examination of the facts as they appear to us.

The phenomenon of the arising and passing away of all things as taught by the Buddha is known as *dependent origination*.[17] The existence of all phenomena depends on the existence of other phenomena. Nothing can have its origin in isolation from anything else. We are born of our parents, with the DNA we have inherited from them, and theirs from others, still under the laws of entropy and change. Cratylus' river cannot exist in the first place—if it ever did!—without the existence of water, or the pathway that it ran down. Indeed, Cratylus would not have had his insightful idea were it not for his teacher, Heraclitus.

We live in a cosmos where there is only change, an endless cycle of flux, arising and passing away. The thingness that we experience is real only in that it exists fleetingly. The things which provide the necessary conditions for anything to arise are also depending on other things.

Despite the huge variation of thought that can be found among all the schools of Buddhism that span Asia, this doctrine is alive in all of them.[18]

Think of the many parts there are in a train engine. All those materials came from something else located elsewhere, that had to be invented and tested. The metals that supply them were fused within stars and had to come to Earth via the impacts of asteroids flung from cosmic explosions far away. And the web of entanglement is not simply made of objects. Although one commentator has said that "Human being is awakened in thingness," the Buddha would say that genuine awakening can only happen when we let go of seeing the thingness around us, and begin to understand things as impermanent instead.[19] In fact, Buddhism would say that it is a web of entanglement of dependent phenomena, all conjoined with impermanence.

Nothing is fixed. But just like some of the laws of geometry, the law of change itself must be changeless. Otherwise, the law would not hold. Because it exists beyond the conditions that it describes, the law of dependent origination is somewhat transcendental.

The Buddha deduced two other characteristics of life on foot of his teaching of dependent origination. One says that because of the law of change, nothing that we want can last. There is nothing which gives any lasting satisfaction. Any good things must end. There is an upside though, being that any unpleasant thing or event shall also pass. But we exist enveloped by this *dukkha*—the original Sanskrit word, which is still used among Buddhists today. *Dukkha* is in all things, and it is often translated as "unsatisfactoriness."

From a kinetic point of view, we can say that all motion, desire, and action arise because of this. It is the starting point and instigator of all human and animal behavior. *Dukkha* is the driving force behind the phenomena of human itchiness, curiosity, and restlessness. We're always looking for another thing or place to go to. We want to improve our quality of life, because there's a little bit of unsatisfactoriness inherent in what we have now.

The other deduction that the Buddha made from the law of change was that there cannot be any such thing as a fixed self. If everything is in process of change, then so are we. The self is a constantly changing entity, he argued. We are clearly not the same people that we were many years ago, so why would we presume that we are the same person as we were even yesterday? Many people find this idea difficult to stomach. And this is understandable when we consider that such a huge part of our culture is invested in "hyper-individuality." The TV shows, newspapers, and product advertisements all appeal to the idea that the "self" is the most important thing in existence. These aspects of modern life are only successful because they play with the *dukkha* in our everyday experience. The denial of self was a spectacular revolution in spiritual thought at the time.[20] The more one contemplates it, the more they must veer towards the reality of Cratylus' river. Although there is a conscious experience, and the existence of memories, that doesn't mean that we are a "thing." We are, like the river, a phenomenon, a process of change, of kinesis.

But the Buddha also offered a way out of this conditionality, and he taught this to his ascetic friends when he returned to Sarnath. He told of escape from this form of existence to another, one of *Nirvana*. The gaining of enlightenment, or to put it into more simple terms, the ridding oneself of any desires, is the path to an alternate state. Part of this path involves the practice of meditation, a witnessing of the activities of the mind with the mind itself, all while remaining physically motionless. In this beautiful passage from a Buddhist text dating to several hundred years after the Buddha, we find a description of the elusive and unintelligible nature of the compassion that an enlightened being has for others, showing the impossibility to understand it with ordinary concepts:

> [It is] like the appearance of matter in an immaterial realm; like a sprout from a rotten seed;
> like a tortoise-hair coat; like the fun of games for one who wishes to die....
> like the existence of desire, hatred, and folly in a saint....
> like the perception of color in one blind from birth....
> like the track of a bird in the sky; like the erection of a eunuch;
> like the pregnancy of a barren woman....
> like dream-visions seen after waking....
> like fire burning without fuel....[21]

The speaker is describing how he sees others in a non-objective way. It is purely sensing, experiencing, without the projection of thingness onto them, something impossible to put into words. They put it into negative terms, as we can only understand what it is not.

And just like early Greek philosophy and Taoism, there is a central static oneness to the cosmos. Nirvana, according to the Buddha, is a place or state where we are free from restlessness. The itch to move is overcome when one experiences enlightenment, hence his standing still in the forest with the bandit. The project of human development is complete and the ultimate stasis is achieved. The Buddha gave both sides of existence in his *Dharma*, the rational way we can understand our ordinary life and also the other one. The latter is deep, cosmic, incalculable, and ungraspable. But it *is attainable*, he tells us. We are shown two parts, one moving and elemental, the other still, whole, integral. Or we can say that one is expressed through an itch to move; the other in the resolution of that urge.

Kinesis drives all of this, as there is only this constant change. Buddhists themselves often talk about their spiritual life in terms of "going forth," as in leaving the life of living in a home, and associations with place, so that they can attain realizations and insights into the nature of themselves and the way life works. The Dharma is described as a "path" to awakening, for those who follow it. Indeed, the three great historical waves of Buddhism are considered as "vehicles," as if they only serve to transfer the seeker to their enlightenment, after which they are redundant.[22]

The Archimedean Point

Not unlike the depiction in modern physics, Buddhism holds that an independently existing world is something that we cannot be absolutely certain of. In terms of kinesis, seeing motion is really relative to the observer. This is as true in Einstein's theory of relativity as it is in the mysterious and mind-bending world of quantum mechanics. It is now an accepted fact that we cannot obtain a position to see the world without interfering with it. This position, traditionally known in philosophy as the *Archimedean Point* was where one could theoretically observe without interference, like the independently existing time-world of the post-railroad nineteenth century, or the position of being outside of time altogether, or of seeing the whole world as if it were on a map.

This objective viewpoint of things is an impossibility says the Buddha. Because of dependent origination, we are by nature interconnected to all things in one way or another. And although we do not have complete control, we do have some. And therefore, we can affect some things around us. We exist within the dimensions of space and time, interacting within a physical context. From the point of view of kinesis there is always an effect from any human motion, however slight we may feel it is. It is reflected in Werner Heisenberg's discovery that electrons exist *in relation* to each other. We cannot know all the information about a particle, its speed, direction, and so on. The more we know about one aspect of it the less we know about the other.[23]

This has been the case in other experiments in quantum physics, in particular some results in which electrons apparently convert themselves into waves and then back again. When examined further by putting more detecting instruments, the electrons seemed to behave as if they "knew" that they were being spied upon. The physicists involved were affecting the particle's behavior by virtue of their observation.[24] We are truly interconnected, argued the Buddha, and as the saying goes, "no man is an island."

Many of the kinetic mysteries that confront us today are indeed mind-boggling. The world of quantum physics is deeply counterintuitive. Particles can behave like waves, and waves like particles. Possibly lying underneath everything is only energy, matter being an illusion. And because of the Buddha's discovery of dependent origination, there is no fixed self, but yet our presence can affect what is observed. The river always changes, and is also not there, and if we weren't here to see it, could we presume it would exist?

In the experience of Nirvana, or enlightenment, there is an ultimate stillness. Attaining this state means to have escaped the restraints of matter and kinesis. The same can be said of the *Tao* and the *Logos.* But in the end, concepts and words fail us and like the other spiritual traditions, it lies beyond our comprehension. Perhaps that is exactly why these ideas can be so enticing. They ignite our innate curiosity, awakening our awareness of *dukkha* and desire. Another elegant passage from the canon of Buddhist literature describes that path which leads to stasis,

> E Ma O
> Dharma Wondrous Strange.
> Profoundest Mystery of the Perfect Ones.
> In the unmoving, all things come and go,
> Yet in that movement, nothing ever moves.[25]

The unmoving *Tao*, *Nirvana*, or *Logos* seem to be a sort of end-point for the dichotomy of the kinesis-stasis model. There is, at the ordinary human level, a lack of *separateness* in the universe. No motion, and no being is existing apart from everything else, our words and actions have consequences, but there nevertheless exists some sort of desire for something beyond this interconnection. Even if Axial thinkers can make many abstract thought experiments with the movement of lines in the perfect spaces of geometry, movement cannot exist by itself.[26] Kinesis depends on objects to move, in order to express itself.

And because we are embodied beings, with our kinesis being felt sensually by the body, and thoughts occupying our minds, it became a cultural norm to consider these two as separate aspects of the human being. Again, it reflects a pair of opposites, just like kinesis and stasis, but this distinction is an "artificial dichotomy" according to Jung. It is an error, he claimed, "which is unquestionably based far more on the peculiarity of intellectual understanding than on the nature of things."[27] There is also a remarkable tendency to associate permanence with our bodies, and change with motion.[28] It is towards resolving this model of a pair of opposites in motion and stasis where we shall conclude.

Chapter 15

Advaita

I am motion in space.[1]

—Edmund Husserl

This book has followed several themes throughout. These reflect fundamental characteristics of human and natural life, facts of our experience on Earth. The first is the Heraclitean flux, the idea that ever-new waters are flowing in the river. The law of change is perhaps the primary characteristic of the universe in which we find ourselves. Secondly, we can *read into* movement to a certain degree. The variety of ways in which humans move and have moved through history can tell us something about their societies and values. We saw this played out in the chapters on bicycles, trains, and automobiles. There is discourse to these modes of kinesis, because they are culturally expressive.

Perhaps the most important and immediate theme is the urge to explore, travel, and act upon our innate curiosity. It's what drives much of human activity and seems to be insatiable. Once an answer to a given problem is found, a new problem is discovered and worked upon. Indeed, scientists are often quoted as saying that when they are less than totally clear on something, they are at their happiest, because they are most deeply engaged with their subject at its most fundamental level.[2]

A fourth theme plays against the kinetic desire. It is a spiritual need for stillness, the overcoming of motion, an urge towards stasis. It is what is described as the *Logos, Tao*, or *Nirvana*. For some, it might be understood as the completion of the human project. It was possibly a major factor in the settling of our nomadic ancestors, and it can be seen to manifest in our modern fixation with ownership, entanglement, and our belief in what I call "thingness." Stasis was the metaphysical aim of Axial spirituality. Compared to contemporary spirituality which is too-often centered on the individual, the Axial religions were social projects, hence the beginnings of utopianism in this era.

These aspects to movement form the underbelly of all that has been discussed in the book thus far. Humans need and desire both kinesis and stasis, to move and to rest freely. The Axial Age revolutions suggest that stasis was attainable through mind. It is the vehicle of transcendence and escape from the endless movement, leading to the ultimate and final motion, the one that brings us to extinguish that itch.

Relocation to the Mind

The importance of mind is reflected in the Buddha's instructions to look into our immediate experience. As he lay dying among his followers, his last words were "strive on untiringly," urging his disciples to remain mindful of the reality of change.[3] It could be said that Gautama was the first psychologist, at least in terms of putting one's own mind in the crosshairs of our attention and examining it rationally. The best-known collection of the Buddha's teachings, the *Dhammapada*, opens with the following lines:

> Experiences are preceded by mind, led by mind and produced by mind. If one speaks or acts with an impure mind suffering follows, even as the cartwheel follows the hoof of the ox drawing the cart.
>
> Experiences are preceded by mind, led by mind and produced by mind. If one speaks or acts with a pure mind happiness follows like a shadow that never departs.[4]

The teaching here is that consciousness is the starting point of everything that occurs in the human experience. It is in the mind where the experiments in farming were born. The minds of men and women were where bicycles, cars, trains, airplanes came from and where the necessary infrastructure for all of those was designed. Our cities come from the human mind, and can be even understood to be maps of consciousness. Utopias also come from the mind, and reflect its nature in their intangible existence. For the practicing Buddhist, these opening lines contain an ethical instruction, suggesting that with mind comes agency, and that our actions have consequences.

The Axial relocation to the mind is a culturally significant development in our story of human kinesis. Thought is physically static, happening in a non-embodied realm. Experientially speaking, thought is the only phenomenon that humans can experience which is physically movement-free, but is always changing nevertheless. This led Axial thinkers to attempt to quiet the mind, through non-action and meditation, and to seek out meditative states of experiential absorption and bliss as if it were a more subtle part of a continuum of the human body.[5] In order to consciously experience stasis, it must be on a different level of consciousness altogether which is why spirituality became involved.

One hundred thousand years ago, our defining characteristic was bipedalism—now it is our capacity for complex thought. But we are precisely the same creatures as those who first wandered the Earth. If one had a time machine, they could take a child from back then and place it anywhere today and it would integrate and behave just like its peers.[6] The differences between us are merely cultural. From the first agricultural societies onwards, simple bodily motion was replaced with its extensions, with horses and wheeled vehicles and over that vast expanse of time the extensions of motion have coincided with the focus of the human project moving away from the body and into the mind.[7]

Feeling

But kinesis is a different phenomenon from thought. We don't *think* movement, we simply feel it physically. If we were to really examine our conscious experiences, we may find that aside from thought, which is merely immaterial and, in a way, unreal, there is only one other mental event. This is the sensation that we feel. Indeed, we could even say that these sensations *are* consciousness itself—because the more we go into it, the more we realize that they can't be separated from our experience. These are also the sensations that we may observe while we move. In fact, it is much more than that. Movement and sensation are inseparable, and experience must come through our senses.[8]

We can also re-affirm our control and agency over our bodies too. A conscious agent can decide to move, say their left arm, and it just happens. There's no decision *how* to do it—all that detail belongs in the sub-conscious part of the mind.[9] So, the relationship between our mental events and physical ones regarding our bodies is perhaps not a relationship between two things at all. Although it's possible to consider it all at a closer level, the mental and physical experiences in that example could be understood to be one and the same.

The great phenomenologist philosopher Maurice Merleau-Ponty rejected the distinction between mind and body that dates all the way back to René Descartes in the sixteenth century, saying that "I am conscious of the world through the medium of my body."[10] He suggested that the human body was a subject, rather than an object, and it is the first port of call, in terms of our experience. The body itself is a form of consciousness, according to Merleau-Ponty, kinesis being crucial to bringing body and mind together. For many centuries, possibly since the emergence of second-order thinking, a duality has existed in the spatialization of our existence. Mind and body had been conceived of as two different things, the latter merely housing the former.

We have seen in some chapters how engaging with kinesis can have certain effects on the mind. One of these examples was the tradition of the running monks in Chapter 3. Another way is through attempting to attain a total bodily stillness. Doing this serves to simplify our mental experience, we can simply sit still as meditators do. One finds that one's sensations become increasingly subtle, and knowledge of one's consciousness deepens. The tradition of meditating—although today most closely associated with Buddhism—was part of all the Axial traditions, and was an important part of the spiritual practice of early Christianity too. It was during the Axial Age that humans began "making themselves aware of the deeper layers of human consciousness."[11] The Axial Age was possibly the restoration of how the nomadic hunter-gatherers were mentally and spiritually.

It marked the beginning of a spiritual search which still carries on today, something based on the intuitive feeling that "there must be something more than just this." It's a product of the spiritual revolution, but also of an age of rational thinking. We have an *intuition* towards stasis. Many people feel that there is some form of stasis beyond what we experience in ordinary daily life. But there is also some association with kinesis—if we remember the many burial sites which contain ships, wagons, and chariots (Chapters 4 and 6). It's as if in the ultimate reality, or even in death itself, kinesis and stasis are not exactly opposites, but form a sort of complementarity.

From Heraclitus' point of view, this way of thinking is the *Logos* manifesting itself in us. And we need to look at the world, and us within it, beyond the simple dual nature that we use so readily. He tells us that "Good and bad are the same."[12] Superficially it is a simple paradox, but he's pointing to a different way of thinking, one beyond duality, where a new reality is formed out of the tension of the supposed contradiction.

Some traditions of Buddhism attempt to alight one's attention on a contradiction for this very reason. Perhaps one of the best-known methods is the use of *kōan* ("riddle") in the Zen practice of Japan (or Ch'an in China).[13] There are many literary examples of teachers giving *kōans* to their students. They are sent away with a rationally impossible puzzle and left to struggle in meditation with it. One of the more famous examples from Medieval China involves a master and student monk examining a bamboo rod, known as a *shippei*:

> You monks, if you call this a shippei, you are adhering to the superficial fact (or negating its essence). If you do not call this a shippei, you are opposing the fact. Tell me, you monks, what will you call this?[14]

Here we are in a tricky situation and left wrestling with an impossibility. The monk has no choice but to sit with and accept the difficulty until something else emerges from beyond their rational mind. Eventually, when reason breaks down, it is said that they gain some form of spiritual insight. With enough of these insights, one passes through a "gateless gate"—a remarkably pithy expression suggesting kinesis (passing through it), stasis (the rational contradiction), thingness (the gate), and the absence of all the above ("gateless").

Their method for seeking stasis was deepening the connection between mind and sensation. By engaging with the opposite of kinesis, people were encouraged to unite stillness with the direct awareness that comes with motion. It reconciles the urge to wander with the urge for a connection in the context of something greater than ourselves. It does so using the realization of Heraclitean flux, as well as the realization of Cratylus' idea of there being no fixed *thingness* to anything, if all things are in flux. That means that things are empty by their nature. They are void of any identity, itself an immaterial product of thought. But kinesis through sensation is a more direct form of reality. Kinesis *is* change. This is why elsewhere we find Heraclitus telling us that "We step and do not step into the same rivers; we are and are not."[15]

Back to Smooth Spaces

To the mind of the modern reader, it may be ironic that at the time of a new spiritual eruption, there came with it the dawn of the modern rational consciousness. *Logos* emerged at the same time as the studies of motion by the early Greek geometrists. At the time of a Taoist cosmic unity of all things, there appeared schools like that of Pythagoras and later Euclid—whose treatises on geometry were the standard for 1,500 years. They developed a new relationship with space, seeking to understand and striate it in the abstract thought of the new rational mind, where beforehand it took place on the land.

The Canadian philosopher Brian Massumi argues that "Before measurement there was air and ground, but not space as we know it."[16] The measurement he describes here is a product of mind, and a relationship with space that comes specifically from the Agricultural Revolution. When the Axial Age arose some 7,000 years afterwards, this measurement of space began to occupy an intangible world driven by second-order thinking. A further irony is the emergence of geometrical space and a spiritual—or smooth—space at the same time. Why would this be? The answer is probably because geometry arose outside of the tangible and material world. It only exists in the purely abstract mindworld powered by elaborate mathematics.

All three Axial traditions suggest the possibility of the existence of a higher consciousness, in which—should we ever attain it—the mind is integral and unified in its stasis. If we compare that state to the types of space, it is a smooth one, a space without striated lines, limitations, or borders. It is sky-like and freeing, not unlike the sea which the Polynesian sailors traversed, or the limitless space that the Apollo astronauts flew through. The space of Axial spirituality produced the idea of a smooth space of stasis, as if reflecting a utopian vision of nomadic space.

But if the Axial thinkers were going back to nomadic values of a lesser engagement with materialism and thingness, they could not do so spatially. Space had been striated by so many agricultural generations living among the straight lines of irrigation, reimagined in the city, that seeing it in this way was deeply embedded in the mind of mankind by that point. Both geometry and land ownership were powerful forces and still are today. And that clinging on to striated space comes from our entanglement and tendency to view the world in only one way, as a "thing."

Hunter-gatherers would have experienced space extremely differently from how we do today, probably as a more fluid continuum, hunting in their meditative silence as discussed in Chapter 3. That smooth space is what the utopias are sometimes reminiscent of. It is a flux, nebulous, endless, and immeasurable. When the emphasis changed to the realm of the intangible, we understood all space the same way we do in a city. Hence it was geometry which marks this new expression of consciousness historically.

Massumi says that "the idea that we live in Euclidean space and in linear time excludes the reality of change."[17] That is, it excludes the Heraclitean flux. There is a difficulty here because the rules of a geometrical world hold true and don't change. But Massumi points to an experiential reality, one where rationalism isn't the only tool in our kit. True reality—to an Axial thinker—is where we unify the rational and spiritual sides of our being, and examine our experience of sensation, of motion, and striving for stasis. The world is more than a series of predictable patterns according to a given set of laws.

Another wonderfully pithy remark is the quote which opens this chapter, from the philosopher who established phenomenology—a branch of modern philosophy which puts the human experience at the center of examination—Edmund Husserl. He claims that he *is* "motion in space." What this means to our story is that his being, or experience, is bound up with kinesis, in the context of a space in which he exists, all these elements inseparable from each other. Many of us consider our own consciousness to exist in a space, somewhat resembling a physical one, perhaps inside of our heads.

Revisiting Time

Like the effect that the train had on time and space by "annihilating" it, as the phrase went, we also spatialize time, as if kinesis gives rise to it, by giving rise to time itself (you may remember that kinesis is primary to time in some ancient and modern philosophy). Many of us think of large spans of time, a century say, as going from left to right. Often, we think of the past as behind us, and the future before us. This is a distinctively Western cultural characteristic. For some non–Western cultures it is the other way around: the past is in front of us, visible, as something we must deal with, and the future behind us, unseen, unknowable. For others, time may even flow uphill.[18] Even time moves.

The more we attend to our experience of mind as it unfolds, the closer we get to Cratylus' river (or lack thereof). Like an atomic physicist looking ever more infinitesimally, finding more and more emptiness, the more attention given over to the waters in the river and the flux of life, the more we're left with everything really being empty of any thingness. Kinesis is the only concrete reality.

We may even question if the self is fixed (as mentioned in Chapter 14) and if there really is a barrier between ourselves and the space that surrounds us. The closer we look, the stranger and less intuitive things usually become, which seems to make up much of the narrative of quantum physics. The further one goes into the matter, the more they have to disentangle some of the big ideas of how the world works.

There are no gaps between things when we look very closely. Rather, we are osmotic bodies, passing through the fluid of an airy atmosphere. And so, there is a "non-duality" to this, or to use the Sanskrit term, *advaita*.[19] We feel real, and separate from the space around us, or the objects that we see, and other people and so on. Most if not all of us do feel like we're an entity, experiencing the world, enjoying our free will and agency. But on closer inspection it's not so simple, like the river, we are and we are not separate.

In fact, when we genuinely consider non-duality, it follows that it must encompass all things, even duality itself. Duality and non-duality are therefore not opposites, but the latter envelopes the former. *Advaita* isn't simply the opposite of duality, it's the transcendence of it. If it were the opposite of duality, then itself would simply form another duality. So non-duality *includes* duality, as if wrapping itself around it. In the same way stasis isn't simply being still, it is the cultural attempt to transcend our embodied attachment to kinesis.

Parmenides—whom we met in Chapter 12—proposed one single *Being* which must include any ideas of "non–Being."[20] His is just one entity, a oneness which is the fundamental characteristic of the cosmos. In the same way, non-duality includes duality. We may call it *Nirvana*, or *Tao*, but what we call it doesn't matter. Indeed, by calling it a name, by even conceiving of it mentally or verbally, we are diminishing its reality, so say the Buddhists and Taoists.[21] Just like the Zen master with the bamboo rod, by labeling it we ignore its true nature and create a duality by associating it with thingness. This is why stasis lies beyond ordinary rationality.

The separation of oneself from others is perhaps the cornerstone upon which to build an agricultural and capitalistic society. Without a strong sense of individual self, there cannot be private property. Over the seven or so millennia between the Agricultural Revolution and the Axial Age, the conditions of modernizing society were becoming ever stronger. After pottery arose, other technologies were developed such as writing, enabling a huge increase in storage and transmission of information, contributing to huge changes in the nature of mind, as memory became outsourced to clay tablets and later paper. Private property also evolved, marking a totally new way of thinking within society, and interacting with others through exchange of commodities. The modern economic unit was established—the nuclear family. The idea of a monogamous man and woman who bring up their own children was a development dating from this time too.[22]

The Axial rejection of materialism and the city can be seen in this context as a push to reconnect with the nomadic life that preceded it by many thousands of years. This is why it is characterized by a move out from the city.[23] It is also why utopianism dates from this time, the fond remembrance of previous eras of happiness and abundance.

All these factors increase the quality of duality in the human consciousness. Both writing and private property worked in tandem with each other with the effect of taking people further away from meditative and mentally equanimous states. Private property is the single most powerful way to bolster the idea of a fixed unchanging self and allows individuals really to see themselves as such. Through our possessions, we can—and too-often do—compare ourselves to others. We can even use our possessions to define who we are. This is what happens in the upper niches of wealthy, consumeristic societies. Today, the culture of using material goods to identify oneself is huge.

This explains how the relationship with space changed through the extensions of motion. Wagons, carts, horse-riding, and modern extensions such as the bicycle and car all take with them an embedded relationship with space, and the other discourses which have been examined throughout this book. Much of what all this points to is that natural kinesis for the nomad may be a non-dual experience of the body which moves. As the photographer Bill Hatcher says, "The idea of connecting to our ancestral past requires us to locomote as we are evolved to do, using our senses and making sure the mind and body are in union."[24]

The Axial thinkers were reacting to the way things had been going since. Their practices of asceticism, moderation, reflection, and meditation were a way of balancing against the ills of an increasingly consumeristic society. The Axial culture encourages a direct sensation of motion of the mind, suggesting that it is this which brings us to a non-dual experience. Most rejected ownership of possessions, seeing an unending emptiness with entangled thingness. Possessing little and being less entangled was a necessary condition to live a life of virtue. And not just that, they attempted to get to a place where movement was no more.

Thusness, Freedom, Dance

The fact of non-duality is a spiritual truth rather than a scientific one. It is something that can be experienced in very deep states of concentrated awareness, something akin to the "flow state" described in Chapter 3. We can get a glimpse of the unity of the motion in all things.

The reality of movement is therefore far more cosmic than we may often think. A "cosmic beat" sounds throughout the universe, existing at every level from the galactic to the infinitesimal.[25] To truly understand movement, we must not only understand stillness, but understand that at yet another level there is no difference between the two. As far as we can see into the cosmos there is no stillness. There is no changelessness. As Isaac Newton pointed out, we can only genuinely witness motion from a fixed point, but that cannot even exist. True kinesis therefore consists of the non-duality of both motion and stillness. One cannot exist without the other, and we're in the thick of it.

Stasis then appears as something outside of kinesis, enveloping it as a non-duality. It is something within which kinesis always must occur, but stasis itself is beyond concepts. It equates to "thusness," an existence which is "free from conceptual elaborations and the subject-object distinction."[26] It manifests itself in us as a type of yearning, as if we are being called to by something outside of our ordinary world. The fact that we can invent—or discover—that law of change, is proof of the relationship we have with stasis.

The urges of stasis and kinesis, the eternal *Tao*, the Buddha's *Dharma*, the automobile, bicycle, the trains and their effect on society at the time—in short all that has been dealt with in these pages—have one greater common theme. They all relate in some way to *freedom*. There is the activity of colonization, going to new lands, or developing a strong spiritual relationship with places, growing one's own food, which are manifestations of this, the freedom to move with bodily agency and the freedom of a different sort gained by being still. Perhaps the most obvious examples come in the form of the bicycle and automobile.

The great political philosopher Thomas Hobbes saw the human condition as one without escape to stasis. We are condemned to our bodies until our death, with only fleeting moments of mental stasis possible from time to time, even though his political ideas mirrored the idea of an overarching ruler, an echo perhaps of Xenophanes or Parmenides but in the political realm. "There is no such thing as perpetual tranquillity of mind," he wrote in 1651, "because life itself is but motion, and can never be without desire."[27] But he nevertheless saw a defining characteristic of freedom in movement. He considered freedom to be a lack of opposition to one's motion, and being able to move is a type of power that we have.[28]

All of this leads to one important species of human movement that I have only briefly mentioned, dance. Dance can be seen as a type of embodied memory, and be considered an expression of spiritual non-duality. The trance-like experiences that can arise from dance appear in hunter-gatherer culture as well as many spiritual traditions.

Presuming that most readers have danced at some point in their lives, you shall be aware that the mental experience of dancing can be different from any other. This could be anything from a simple "letting go" in order to feel comfortable to a state of trance-like ecstasy. It's a very pleasant experience, to move one's body without a utilitarian reason, in the context of a society which puts our use-value before everything else. Although dancing can often be for a specific purpose—such as causing it to rain, or to lure a lover—it doesn't always need to be so, and the difference may be unclear. Dancing purposelessly, not unlike *dérive*, flies in the face of usefulness, and it can liberate us from modern society's insistence that we need to justify our existence.

In dancing, we move but don't go anywhere, just existing in this dazzling display of pure embodiment, as if mimicking Shiva, the cosmic dancer. To dance is to celebrate having a body, conjoined with the flowing mental experience that comes with it, and has long associations in spiritual traditions.[29] Losing oneself in its ecstasy, the dancer travels beyond their own world of self and other, the word ecstasy literally meaning "standing outside of oneself."[30]

Not going anywhere, it is kinesis without locomotion, and without extension. But it brings with it the transcendental element that so many people before us sought. The dancer, choreographer, and student of human movement Rudolph von Laban considered this outwardly when he stated that a "dancer directly expresses the essence of the world."[31]

Indeed, it may come as no surprise that dancing was held to be one of the greatest arts during the Axial period.[32] Dancing is said to have "awakened and preserved in the soul the sentiment of harmony and proportion. It embraced all the parts of gesture or action. In the times of Plato, the art of dancing had, among the Greeks, such a character of nobility, of perfection, of even utility, as nowadays, is no longer found in it or allowed to it."[33] So, perhaps we have been in decline since the Axial Age, speaking in terms of bodily kinesis and spirituality.

Dancing is not an activity exclusive to humans either. Going all the way back before the emergence of *Homo sapiens*, we see a connection to our closest cousins on planet Earth, the chimpanzees. The apes from whom we split about six to seven million years ago have been documented to enjoy a dance every now and again. Chimps normally dance near water—as if celebrating the first conditions for the existence of life itself. This happens mostly through periods of rainfall, but they are also known to dance in waterfalls. These waterfall dances are described by the famous primatologist Jane Goodall,

> His pace quickens, his hair becomes fully erect, and upon reaching the stream he may perform a magnificent display close to the foot of the falls. Standing upright, he sways rhythmically from foot to foot, stamping in the shallow, rushing water, picking up and hurling great rocks.[34]

It would seem, then, that the chimpanzee has the same relationship with dancing as ourselves. Dancing is so primal and spiritual a thing that all human cultures have had dance as part of their life. Like music, there isn't a known society without dance, so it's deductible that somehow, albeit mysteriously, it is essential to our survival. Perhaps the phenomenon of dance is related to the seed of human kinesis at the deepest level.

Dance is motion, rhythmic motion, a tiny sample of the cosmic beat. It is the strange, illogical, mysterious, and wonderful combination of human motion with music. Dance historian Curt Sachs explains that "dance is the mother of the arts. Music and poetry exist in time; painting and architecture in space. But the dance lives at once in time and space."[35] Dance is inseparable from those who do it, embodying thusness, a form of kinetic expression that does not reduce time and space, but unites them. We can only witness it happen, words no longer serving us from that point onwards.

In a world where even the future of movement is unknown, perhaps the only thing left for us is to dance. And what better way to see its nonduality than with the words of William Butler Yeats,

> O body swayed to music, O brightening glance,
> How can we know the dancer from the dance?[36]

Chapter Notes

Chapter 1

1. Heraclitus, *Fragment 51*, in Charles H. Kahn, *The Art and Thought of Heraclitus: A New Arrangement and Translation of the Fragments with Literary and Philosophical Commentary* (Cambridge: Cambridge University Press, 1979), 871, Kindle.
2. Attributed to Cratylus in Aristotle, *Metaphysics* IV, 5 in *Aristotle: The Complete Works of Aristotle*, trans. Jonathan Barnes (Oxford: Oxford University Press, 1991), 3032, Kindle.
3. Michel de Certeau, *The Practice of Everyday Life* (Berkeley, CA: University of California Press, 1984), 115.
4. John Urry, *Mobilities* (Cambridge: Polity Press, 2007), 13.
5. Although little is known of Cratylus, Diogenes Laertius describes him as a follower of Heraclitus, in *Lives of the Eminent Philosophers* Book III/6, trans. Pamela Mensch (Oxford: Oxford University Press, 2018), 136.
6. Adam G. Riess et al., "Observational Evidence from Supernovae for an Accelerating Universe and a Cosmological Constant," *The Astronomical Journal* 116, no. 3 (1998): 1009–38.
7. The first European suggestion of "atomism" is thought to have been by Leucippus followed by Democritus c.460–c.370, approximately three generations after Heraclitus. See Carlo Rovelli, *Reality is Not What it Seems: The Journey to Quantum Gravity* (London: Penguin, 2016), 130–690, Kindle.
8. Jiddu Krishnamurti, *Krishnamurti's Notebook* (London: Harper & Row, 1976), 39.
9. Thomas Nail, *Being and Motion* (New York: Oxford University Press, 2019), 29.
10. Henri Bergson, *The Creative Mind: An Introduction to Metaphysics*, trans. T.E. Hulme (New York: Putnam's Sons, 1912), 1944, Kindle.
11. Karl Marx and Friedrich Engels, *The Communist Manifesto* (London: Penguin Books, 1967), 223.
12. See Jeremy Black, *Visions of the World: A History of Maps* (London: Mitchell Beazley, 2003).
13. Brian Massumi, *Parables for the Virtual: Movement, Affect, Sensation* (Durham: Duke University Press, 2002), 1, Emphasis in original.

Chapter 2

1. Rebecca Solnit, *Wanderlust: A History Of Walking* (New York: Penguin, 2001), 669. Kindle.
2. Alice Roberts, *The Incredible Human Journey: The Story of How we Colonised the Planet* (London: Bloomsbury, 2009), 76, Kindle.
3. Christopher Seddon, *Humans: from the Beginning: From the First Apes to the First Cities* (London: Glanville Publications, 2014), 36.
4. Carl Hall, "Walk Before You Talk," *Science* 292, no. 5526 (Jun 2001): 2429. Joseph Amato, *On Foot: A History of Walking* (New York: New York University Press, 2004), 22.
5. James Earls, *Born to Walk: Myofascial Efficiency and the Body in Movement* (Chichester: Lotus Publishing, 2014), 378, Kindle.

6. John Napier, "The Antiquity of Human Walking," *Scientific American* 216, no. 4 (Apr 1967): 468.

7. Earls, *Born to Walk*, 278.

8. Earls, *Born to Walk*, 227.

9. See Caroll Ward et al., "The new hominid species Australopithecus anamensis," *Evolutionary Anthropology: Issues, News, and Reviews* 7, no. 6 (1999): 197–205.

10. Richard Wrangham, *Catching Fire: How Cooking Made Us Human* (London: Profile Books Ltd., 2009).

11. Ian McDougall et al., "Stratigraphic placement and age of modern humans from Kibish, Ethiopia," *Nature* 433 (2005): 733–6.

12. William H. McNeill, *Keeping Together in Time: Dance and Drill in Human History* (Cambridge, MA: Harvard University press, 1995), 14, Kindle.

13. McNeill, *Keeping Together*, 10.

14. Judith Lynne Hanna, "To Dance is Human," in *The Anthropology of the Body*, ed. John Blacking (London: Academic Press, 1977), 222.

15. Hanna, "To Dance is Human," 225.

16. McNeill, *Keeping Together*, 42–8.

17. McNeill, *Keeping Together*, 152.

18. Tim Robinson, *My Time in Space* (Dublin: Lilliput Press, 2001), 103.

19. See Theodor Strehlow, *Songs of Central Australia* (Sydney: Angus and Robertson, 1978). Also Bruce Chatwin, *The Songlines* (New York: Penguin, 1987).

20. Amato, *On Foot*, 19.

21. K.M. Lucibello et al., "Examining a training effect on the state anxiety response to an acute bout of exercise in low and high anxious individuals," *Journal of Affective Disorders* 247 (Dec 2019): 29–35.

22. Jean-Jacques Rousseau, *The Confessions* (Feedbooks, 1768), 262, Kindle.

23. Amato, *On Foot*, 53.

24. General Authority for Statistics, Kingdom of Saudi Arabia, "Hajj Statistics: 2019–1440," accessed Feb 2, 2023. https://www.stats.gov.sa/sites/default/files/haj_40_en.pdf.

25. Solnit, *Wanderlust*, 107.

26. Lauren Artress, *Walking a Sacred Path: Rediscovering the Labyrinth as a Spiritual Practice* (New York: Riverhead Books, 2006), 23–44.

27. Daniele S. Lizier et al., "Effects of Reflective Labyrinth Walking Assessed Using a Questionnaire," *Medicines* 5, no. 4 (2018): 111.

28. Geoff Nicholson, *The Lost Art of Walking: The History, Science, Philosophy, and Literature of Pedestrianism* (London, Riverhead Books, 2008), 183.

29. Penelope Reed Doob, *The Idea of the Labyrinth: from Classical Antiquity to the Middle Ages* (Ithaca, NY: Cornell University Press, 1992), 17.

30. Jeff Saward, *Labyrinths & Mazes: A Complete Guide to Magical Paths of the World* (Asheville, NC: Lark Books, 2003), 60–70.

31. Curt Sachs, *World History of the Dance*, trans. Bessie Schönberg (New York: W. W. Norton & Company, 1937), 151–2.

32. Mary Bond, *The New Rules of Posture: How to Sit, Stand, and Move in the Modern World* (Rochester, VT: Healing Arts Press, 2007), 7. Svenja Lohner, "Sensing with your Feet!" *Scientific American*, July 20, 2017, accessed Jan 20, 2023. https://www.scientificamerican.com/article/sensing-with-your-feet/. Emily Splichal, "The effect of sensory stimulation on movement accuracy," *Lermagazine*, March 2018, accessed Jan 20, 2023. https://lermagazine.com/article/the-effect-of-sensory-stimulation-on-movement-accuracy#:~:text=The%20foot%20is%201%20of,adapting%20type%202%20(FA2).

33. Erik Trinkaus and Hong Shang, "Anatomical evidence for the antiquity of human footwear: Tianyuan and Sunghir," *Journal of Archaeological Science* 35, no. 7 (2008), 1928–33.

34. Marquita Volken, "Arming Shoes of the Fifteenth Century," *Acta Periodica Duellatorum* 5, no. 2 (2017): 25.

35. Tony Waldron, *Shadows in the Soil: Human Bones & Archaeology* (London: Tempus, 2001), 60.

36. Amato, *On Foot*, 6.

37. Urry, *Mobilities*, 64.

38. Tansy, E. Hoskins, *Foot Work: What your Shoes Tell you about Globalisation* (London: Weidenfeld & Nicolson, 2020), 334–713, Kindle.

39. Hoskins, *Foot Work*, 30.

40. Elizabeth Semmelhack quoted in Hoskins, *Foot Work*, 30.

41. Hoskins, *Foot Work*, 595–700.

42. M.H. Kim et al., "Reducing the frequency of wearing high-heeled shoes and increasing ankle strength can prevent ankle

injury in women," *International Journal of Clinical Practice* 69, vol. 8 (Jun 2015): 909–10. Maxwell S. Barnish and Jean Barnish, "High-heeled shoes and musculoskeletal injuries: a narrative systematic review," *BMJ Open* 6, no. 1. (Jan 2016): 1–8. Jacob Barrera Loredo et al., "Influence of High Heels on Walking Motion: Gait Analysis," *Journal of Applied Biomechanics*, Dec 2015. Daniel Lieberman, *The Story of the Human Body: Evolution, Health, and Disease* (New York: Pantheon Books, 2013), 348.

43. Elizabeth Semmelhack, *Shoes: The Meaning of Style* (London: Reaktion Books, 2017), 161.

Chapter 3

1. George Sheehan, *Running & Being: The Total Experience* (Emmaus, PA: Rodale, 1978), 379.

2. Body mass index (BMI) is calculated as weight in kilograms divided by height in meters squared, the idea being that the weight of a given individual will be appropriate for their natural size.

3. World Health Organization, "Physical Activity Fact Sheet," October 2022, accessed Jan 26, 2023. https://www.who.int/news-room/fact-sheets/detail/physical-activity.

4. Amber Sayer, "How many people have run a marathon: World Statistics," *marathonhandbook*, Nov 18, 2022, accessed July 10, 2023. https://marathonhandbook.com/how-many-people-have-run-a-marathon/#:~:text=According%20to%20the%20US%20Census,population%20has%20run%20a%20marathon.

5. Daniel Lieberman et al., "The evolution of endurance running and the tyranny of ethnography: A reply to Pickering and Bunn (2007)," *Journal of Human Evolution* 53 (2007): 439–42.

6. N.C.C. Sharp, "Timed running speed of a cheetah (Acinonyx jubatus)," *Journal of Zoology* 241, no. 3 (Mar 1997): 493–4.

7. Sayer, "World Statistics."

8. Clive Gamble, *Timewalkers: The Prehistory of Global Colonization* (Phoenix: Alan Sutton Publishing, 1993), 95.

9. Campbell Rolian et al., "Walking, Running and the Evolution of Short Toes in Humans," *Journal of Experimental Biology* 212, no. 5 (2009): 713–21.

10. Eduard Alentorn-Geli et al., "The Association of Recreational and Competitive Running With Hip and Knee Osteoarthritis: A Systematic Review and Meta-analysis," *Journal of Orthopaedic & Sports Physical Therapy* 47, no. 6 (Jun 2017): 373–90.

11. Ross H. Miller et al., "Why don't most runners get knee osteoarthritis? A case for per-unit-distance loads," *Medicine & Science in Sports & Exercise* 46, no. 3 (Mar 2014): 572–9.

12. Laura Maria Horga et al., "Can marathon running improve knee damage of middle-aged adults? A prospective cohort study," *BMJ Open Sports & Exercise Medicine* 16, no. 5 (Oct 2019).

13. Ross. H. Miller and Rebecca. L. Krupenevich, "Medial knee cartilage is unlikely to withstand a lifetime of running without positive adaptation: a theoretical biomechanical model of failure phenomena," *PeerJ* 8 (2020): 1–27.

14. Roberts, *Human Journey*, 610–23.

15. Dennis Bramble and Daniel Lieberman, "Endurance Running and the Evolution of *Homo*," *Nature* 432, no. 7015 (Nov 2004): 348.

16. Bramble and Lieberman, "Endurance Running," 345.

17. See Desmond Morris, *The Naked Ape: A Zoologist's Study of the Human Animal* (London: Jonathan Cape, 1967).

18. Charles Darwin, *The Descent of Man, and Selection in Relation to Sex* (London: John Murray, 1871), 166, Kindle. Gamble, *Timewalkers*, 29–30. Stephen Oppenheimer, *Out of Eden: The Peopling of the World* (London: Robinson, 2003), 460, Kindle.

19. D.M. Bramble and F.A. Jenkins, Jr., "Mammalian Locomotor-Respiratory Integration: Implications for Diaphragmatic and Pulmonary Design," *Science* 262, no. 5131 (Oct 1993): 196–7.

20. Louis Liebenberg, "The Relevance of Persistence Hunting to Human Evolution," *Journal of Human Evolution* 55, no. 6 (2008): 1156–9.

21. Earls, *Born to Walk*, 486.

22. Seddon, *Humans*, 64.

23. Daniel Schmitt et al., "Experimental Evidence Concerning Spear Use in Neanderthals and Early Modern Humans," *Journal of Archaeological Science* 30 (2003): 103–14. Chris Stringer, *Lone Survivors: How*

we came to be the only Humans on Earth* (New York: Times Books, 2012), 158.

24. Oppenheimer, *Out of Eden*, 835.

25. Roberts, *Human Journey*, 643.

26. Silke Felton and Heike Becke, *A Gender Perspective on the Status of the San in Southern Africa* (Windhoek: Legal Assistance Centre, 2001), 25.

27. Marjorie Shostak, *Nisa: The Life and Words of a !Kung Woman* (Cambridge, MA: Harvard University Press, 1981), 12.

28. Patricia Draper, "!Kung Women: Contrasts in Sexual Egalitarianism in Foraging and Sedentary Contexts," in *Toward an Anthropology of Women*, ed. R.R. Reiter (New York: Monthly Review Press, 1975), 77–109.

29. Stringer, *Lone Survivors*, 161.

30. Andrew J. Noss and Barry S. Hewlett, "The Contexts of Female Hunting in Central Africa," *American Anthropologist* 103, no. 4 (2001): 1024–40.

31. Randall Haas et al., "Female hunters of the early Americas," *Science Advances* 6, 45 (2020).

32. Paul Ronto, "The State of Ultra Running 2020," Run Repeat, accessed Feb 6, 2023. https://runrepeat.com/state-of-ultra-running. September 21, 2021.

33. Jen Jacobs Andersen, "Research: Women are Better Runners than Men," Run Repeat, accessed 6 Feb, 2023. https://runrepeat.com/research-women-are-better-runners-than-men.

34. Stringer, *Lone Survivors*, 36–7.

35. BBC Earth, "The Intense 8 hour hunt," accessed Jan 27, 2023. https://www.youtube.com/watch?v=826HMLoiE_o.

36. Kenneth E. Callen, "Mental and emotional aspects of long-distance running," *Psychosomatics* 24, no. 2 (1983): 133–51.

37. Bramble and Lieberman, "Endurance Running," 435–6. Roberts, *Human Journey*, 639.

38. Earls, *Born to Walk*, 683.

39. J. D. Lewis-Williams, "Quanto?: The Issue of 'Many Meanings' in Southern African San Rock Art Research," *The South African Archaeological Bulletin* 53, no. 168 (1998): 86–97. Shostak, *Nisa*, 9.

40. Francesco D'Errico et al., "Early evidence of San material culture represented by organic artifacts from Border Cave, South Africa," *PNAS* 109, no. 33 (2012): 13214–9.

41. Benjamin R. Smith quoted in "In South Africa, Discovering the World's Oldest Drawing," The Andrew W. Mellon Foundation, accessed January 27, 2023. https://web.archive.org/web/20200309004930/https://mellon.org/shared-experiences-blog/archaeologists-discovery-worlds-oldest-drawing-highlights-strong-interest-ancient-rock-art/#top.

42. Roger Bannister, *The First Four Minutes* (Stroud, UK: Sutton Publishing Ltd., 2004), 9.

43. Mihaly Csikszentmihalyi, *Flow: The Psychology of Optimal Experience* (New York: Harper Perennial, 2008), 96–100.

44. Csikszentmihalyi, *Flow*, 67.

45. Bernd Heinrich, *Why We Run: A Natural History* (New York: HarperCollins 2001), 12.

46. John Stevens, *The Marathon Monks of Mount Hiei* (Boston: Shambala, 1988).

Chapter 4

1. T.S. Eliot, *Four Quartets* (New York: Harvest, 1943).

2. J. Iriarte et al., "Geometry by Design: Contribution of Lidar to the Understanding of Settlement Patterns of the Mound Villages in SW Amazonia," *Journal of Computer Applications in Archaeology* 3, no. 1 (2020): 151–69.

3. Morris, *Naked Ape*, 93.

4. The famous quote is from the *Apology*, 38a, "life without this sort of examination is not worth living," in Plato, *The Last Days of Socrates: Euthyphro, Apology, Crito, Phaedo*, trans. H. Tredennick and H. Tarrant (London: Penguin Books, 1993), 63.

5. Chao Huang et al., "Evidence of Fire Use by Homo erectus pekinensis: An XRD Study of Archaeological Bones From Zhoukoudian Locality 1, China," *Frontiers in Earth Science* 9 (2022).

6. Quoted in Veronica Greenwood, "Beyond," *Aeon*, Aug 12, 2015, accessed Feb 11, 2023. https://aeon.co/essays/what-drives-the-urge-to-explore-the-farthest-human-reaches.

7. Edward O. Wilson, *The Future of Life* (New York: Vintage, 2003), 3–21.

8. Peter Bellwood, *First Migrants: Ancient Migration in Global Perspective* (Chichester: John Wiley & Sons, 2013), 41–7.

9. Marta Melé et al., "The Genographic Consortium, Recombination Gives a New Insight in the Effective Population Size and the History of the Old World Human Populations," *Molecular Biology and Evolution* 29, no. 1 (Jan 2012): 25–30.

10. T. Goebel et al., "The late Pleistocene dispersal of modern humans in the Americas," *Science* 319, no. 5869 (Mar 2008): 1497–1502.

11. Darwin, *Descent of Man*, 199.

12. Rebecca Cann et al., "Mitochondrial DNA and Human Evolution," *Nature* 325 (1987): 31–6.

13. Eva K.F. Chan, et al., "Human origins in a southern African palaeo-wetland and first migrations," *Nature* 575 (Nov 2019): 185–201.

14. M. Karmin et al., "A recent bottleneck of Y chromosome diversity coincides with a global change in culture," *Genome Research* 25, no. 4 (Apr 2015): 459–66.

15. Russell Thomson et al., "Recent Common Ancestry of Human Y Chromosomes: Evidence from DNA Sequence Data," *Proceedings of the National Academy of the Sciences of the United States of America* 97, no. 13 (June 2000): 7360–5.

16. Chris Stringer and P. Andrews, "Genetic and Fossil Evidence for the Origin of Modern Humans," *Science* 239, no. 4845 (Mar 1988): 1263–8.

17. The geneticist who calculated this is Professor Sarah A. Tishkoff at the University of Pennsylvania. Nicholas Wade, *Before the Dawn: Recovering the Lost History of Our Ancestors* (London: Penguin, 2006), 81.

18. Sarah A. Tishkoff et al., "The Genetic Structure and History of Africans and African Americans," *Science* 325, no. 5930 (2009): 1035–44.

19. Yu Ning et al., "Larger Genetic Differences Within Africans Than Between Africans and Eurasians," *Genetics* 161, no. 1 (May 2002): 269–74. Deepti Gurdasani et al., "The African Genome Variation Project Shapes Medical Genetics in Africa," *Nature* 517 (2015): 327. Bellwood, *First Migrants*, 40.

20. Michael Marshall, "Supervolcano eruptions may not be so deadly after all," *New Scientist*, Apr 29, 2013, accessed Jun 14, 2023. https://www.newscientist.com/article/dn23458-supervolcano-eruptions-may-not-be-so-deadly-after-all/.

21. Alan H. Simmons, *Stone Age Sailors: Prehistoric Seafaring in the Mediterranean* (Walnut Creek, CA: Left Coast Press, 2014), 27.

22. Russell L. Ciochon, and O. Frank Huffman, "Java Man," in *Encyclopedia of Global Archaeology*, ed. Claire Smith (New York: Springer International Publishing, 2020), 4182–8.

23. Simmons, *Stone Age Sailors*, 26.

24. Lady Gregory, *Visions and Beliefs in the West of Ireland* (Toronto: Colin Smythe Ltd., 1976), 15–30.

25. Geoffrey Irwin, *The Prehistoric Exploration and Colonisation of the Pacific* (Cambridge: Cambridge University Press, 1992), 3–6.

26. This compares starkly to the average ratio of one language to each million of world inhabitants. J. Edward. Chamberlin, *Island: How Islands Transform the World* (London: Elliott & Thompson, 2013), 202.

27. Chamberlin, *Island*, 202–4. Jon M. Erlandson and Scott M. Fitzpatrick, "Oceans, Islands, and Coasts: Current Perspectives on the Role of the Sea in Human Prehistory," *Journal of Island & Coastal Archaeology* 1, no. 1 (Feb 2006): 5–32. Bob Quinn, *The Atlantean Irish: Ireland's Oriental and Maritime Heritage* (Dublin: Lilliput Press, 2005).

28. Chamberlin, *Island*, 36.

29. Chamberlin, *Island*, 41–3.

30. Irwin, *Prehistoric Exploration*, 47.

31. Thomas Gladwin, *East is a Big Bird: Navigation and Logic on Puluwat Atoll* (Cambridge, MA: Harvard University Press, 1970), 225.

32. I.C. Campbell, "The Lateen Sail in World History," *Journal of World History* 6, no. 1 (1995): 14.

33. Gladwin, *East is a Big Bird*, 133–73. Irwin, *Prehistoric Exploration*, 7.

34. Nancy Jenkins, *The Boat Beneath the Pyramid: King Cheop's Royal Ship* (New York: Holt, Rinehart and Winston, 1980), 8.

35. Jenkins, *Boat Beneath the Pyramid*, 108.

36. See Eric Powell, "Oldest Egyptian Funerary Boat," *Archaeology: A Publication of the Archaeology Institute of America*, accessed Feb 3, 2023. https://www.archaeology.org/issues/61-1301/features/274-top-10-2012-abu-rawash-funerary-boat.

37. Owen Jarus, "Viking Ship and

Cemetery Found Buried in Norway," *Livescience*, accessed Feb 3, 2023. https://www.livescience.com/63829-viking-ship-cemetery.html. P. Holck, "The Oseberg ship burial, Norway: New thoughts on the skeletons from the grave mound," *European Journal of Archaeology* 9, no. 2–3 (2006): 185–210.

38. Wallace J. Nichols, *Blue Mind: The Surprising Science That Shows How Being Near, In, On, or Under Water Can Make You Happier, Healthier, More Connected, and Better at What You Do* (London: Abacus, 2018).

39. Michel Foucault, "Of Other Spaces, Heterotopias," *Architecture/Mouvement/Continuité* 5 (1994): 49.

40. Foucault, "Of Other Spaces," 49.

41. Bellwood, *First Migrants*, 6.

42. Gamble, *Timewalkers*, 95.

43. Guillermo Altares, "Una humanidad vagabunda: en el futuro, todos volveremos a ser nómadas," *El País*, Nov 24, 2022. Jean-Paul Demoule, *Homo Migrans: De la Sortie d'Afrique au Grand Confinemont* (Paris: Payot, 2022). Gaia Vince, *Nomad Century: How to Survive the Climate Upheaval* (London: Allen Lane, 2022). Anthony Sattin, *Nomads: The Wanderers Who Shaped our World* (London: John Murray Publishers Ltd., 2022).

44. United Nations, "Overcoming Barriers: Human Mobility and Development," United Nations Human Development Report 2009, accessed Feb 8, 2023. http://oppenheimer. mcgill.ca/IMG/pdf/HDR_2009_EN_Complete.pdf. World Health Organization, "World Migration Report 2010," accessed Feb 8, 2023. https://www. iom.int/world-migration-report-2010).

45. Thomas Nail, *The Figure of the Migrant* (Stanford, CA: Stanford University Press, 2015), 184, Kindle.

46. For more on this historical view of the centrality of communication, see Yuval Noah Harari, *Sapiens: A Brief History of Humankind* (London: Vintage, 2011). Kindle.

Chapter 5

1. The original title of Gauguin's painting is "D'où venons-nous? Que sommes-nous? Où allons-nous?" dating from 1897–8.

2. For more on the settling of Madagascar see Mack, *Sea*, 7–10.

3. Aristotle puts motion as a primary and time as secondary. "Time is a measure of motion, and of being moved," *Physics Book IV*, 220b33, in *Complete Works* Vol 1, 73.

4. Ofer Bar-Yosef, "Climatic Fluctuations and Early Farming in West and East Asia," *Current Anthropology* 52, no. S4 (2011): S175–93.

5. Quoted in Wade, *Before the Dawn*, 126.

6. Michael Balter, "The Tangled Roots of Agriculture," *Science* 327, no. 5964 (2010): 404–6.

7. Jacques Cauvin, *The Birth of the Gods and the Origin of Agriculture* (Cambridge: Cambridge University Press, 2002), 7.

8. Ben Shaw et al., "Emergence of a Neolithic in highland New Guinea by 5000 to 4000 years ago," *Science Advances* 6, no. 13 (2020), accessed Feb 23, 2023. https://www.science.org/doi/10.1126/sciadv.aay4573.

9. Sally L. Dillon et al., "Domestication to Crop Improvement: Genetic Resources for *Sorghum* and *Saccharum* (Andropogoneae)," *Annals of Botany* 100, no. 5 (Oct 2007): 975–89.

10. Colin Renfrew, "The Sapient Paradox: Social Interaction as a Foundation of Mind," Nov 14, 2016, accessed Feb 23, 2023. https://www.youtube.com/watch?v=Xgl7b02Ub6Y&list=PLUmOb2SQEIw1fEb6KEP7dy6T-37YqPN45&index=5.

11. Rabindra N. Chakraborty, "Sharing Culture and Resource Conservation in Hunter-Gatherer Societies," *Oxford Economic Papers* 59, no. 1 (Jan 2007): 63–88. James Woodburn, "Egalitarian Societies," *Man* 17, no. 3 (1982): 431–51. Frank W. Marlowe, "Hunter-Gatherers and Human Evolution," *Evolutionary Anthropology* 14, no. 2 (Apr 2005): 54–67.

12. Ian Hodder, "Çatalhöyük: The Leopard Changes its Spots. A summary of recent work," *Anatolian Studies* 64 (2014): 1–22.

13. Harari, *Sapiens*, 2014–6.

14. Trevor Watkins, "Monumentality in Neolithic southwest Asia: making memory in time and space," in *Momentualising Life in the Neolithic: Narratives of Change and Continuity*, ed. Anne Birgitte Gebauer et al. (Oxford: Oxbow Books, 2020), 19.

15. Claudia Sagona, *The Archaeology of Malta* (New York: Cambridge University Press, 2015), 47.
16. Elif Batuman, "The Sanctuary: The World's Oldest Temple and the Dawn of Civilization," *The New Yorker*, Dec 19, 2011.
17. Andrew Curry, "Göbekli Tepe: The World's First Temple?" *Smithsonian Magazine*, Nov 2008, accessed Feb 23, 2023. https://www.smithsonianmag.com/history/gobekli-tepe-the-worlds-first-temple-83613665/.
18. M. Seyfzadeh and R. Schoch, "World's First Known Written Word at Göbekli Tepe on T-Shaped Pillar 18 Means God," *Archaeological Discovery* 7, no. 2 (Apr 2019): 31–53.
19. Watkins, "Monumentality in Neolithic," 22–3.
20. Michael Balter, "Why Settle Down? The Mystery of Communities," *Science* 282, no. 5393 (Nov 1998): 1442–5.
21. Joachim Radkau, *Nature and Power: A Global History of the Environment* (Cambridge: Cambridge University Press, 2002), 64–5.
22. George R. Milner, "Early agriculture's toll on human health," *PNAS* 116, no. 28 (Jul 2019): 13721–3. Marshall Sahlins, *Stone Age Economics* (Chicago: Aldine-Atherton Inc., 1972), 82–92. Jared Diamond, "The Worst Mistake in the History of the Human Race," *Discover Magazine*, May 1, 1999, accessed Feb 20, 2023. https://www.discovermagazine.com/planet-earth/the-worst-mistake-in-the-history-of-the-human-race.
23. Balter, "Why Settle Down?," 1442.
24. Robert McCormick Adams, *Heartland of Cities: Surveys of Ancient Settlement and Land Use on the Central Floodplain of the Euphrates* (Chicago: University of Chicago Press, 1981), 138. Victor J. Katz, *A History of Mathematics: An Introduction* (Boston: Addison-Wesley, 2009), 19.
25. John Reader, *Cities* (London: Vintage, 2005), 31.
26. Marlies Heinz, "Public buildings, palaces and temples," in *The Sumerian World*, ed. Harriet Crawford (Oxon, UK: Routledge, 2013), 180.
27. Guillermo Algaze, "The end of prehistory and the Urusk period," in Crawford, ed., *Sumerian World*, 73.
28. Reader, *Cities*, 25–31.
29. Ian Hodder, *Entangled: An Archaeology of the Relationships between Humans and Things* (Chichester: Wiley Blackwell, 2012), 195.
30. Hodder, *Entangled*, 202. Renfrew, Colin: "Symbol before concept: material engagement and the early development of society," in *Archaeological Theory Today*, ed. Ian Hodder (London: Blackwell, 2001), 128.
31. Hodder, *Entangled*, 87. Harari makes a similar point of humans being domesticated by wheat, and not the other way around. *Sapiens*, 90.
32. Ugo Bardi et al., "Toward a General Theory of Societal Collapse: A Biophysical Examination of Tainter's Model of the Diminishing Returns of Complexity," *BioPhysical Economics and Resource Quality* 4, no. 3 (2019).
33. Edward Gibbon, *The Decline and Fall of the Roman Empire*, ed. Antony Lentin and Brian Norman (Ware, UK: Wordsworth Editions, 1998), 1075–9.
34. Charles Darwin, *On the Origin of Species by Means of Natural Selection / or the Preservation of Favoured Races in the Struggle for Life.* (2nd edition), 6442–5, Kindle.
35. Urry, *Mobilities*, 51.
36. See Marshall McLuhan, *Understanding Media: The Extensions of Man* (Cambridge, MA: MIT Press, 1994).
37. J.C. Margueron, "The Kingdom of Mari," in Crawford, ed., *Sumerian World*, 521.
38. Samuel Noah Kramer, *The Sumerians: Their History, Culture, and Character* (Chicago: University of Chicago Press, 1963), 4, 290. Jacob Bronowski, *The Ascent of Man* (London: BBC Books, 1973), 63.
39. Kramer, *Sumerians*, 103.
40. Jobst Brandt, *The Bicycle Wheel* (Menlo Park, CA: Avocet Inc., 1983), 4.
41. Stephan Lindner, "Chariots in the Eurasian Steppe: a Bayesian approach to the emergence of horse-drawn transport in the early second millennium BC," *Antiquity* 94, no. 374 (2020): 367.
42. Karen Rhea Nemet-Nejat, *Daily Life in Ancient Mesopotamia* (Westport, CT: Greenwood Press, 1998), 182–5.
43. Nemet-Nejat, *Ancient Mesopotamia*, 88.
44. Lewis Mumford, *Technics & Civilization* (Chicago: University of Chicago Press, 2010), 32.

45. This is the Greek "Antikythera Mechanism," discovered in 1902.

46. Alfred S. Posamentier and Ingmar Lehmann, *Pi: A Biography of the World's Most Mysterious Number* (New York: Prometheus Books, 2004).

47. Katz, *History of Mathematics*, 16.

48. Giorgio Vasari, *The Lives of the Artists*, trans. Julia Conaway Bondanella and Peter Bondanella (Oxford: Oxford University Press, 1991), 22–3.

49. Arthur W.J. G Ord-Hume, *Perpetual Motion: The History of an Obsession* (New York: St. Martin's Press, 1977), 41–4.

Chapter 6

1. Attributed to Genghis Khan.

2. Marc Azéma and Florent Rivère, "Animation in Paleolithic art: a pre-echo of cinema," *Antiquity* 86 (2012): 316–20.

3. The importance of this animal in the Chauvet cave paintings is not of singular occurrence. It is also exemplified in other items from this epoch such as the Vogelherd Horse, an ivory carving dating back 35,000 years. This was found in a cave in Germany.

4. Nirwan Ahmad Arsuka, "A Tale of Prehistoric Horses in South Sulawesi," *The Jakarta Post*, Dec 8, 2015.

5. Wade, *Before the Dawn*, 49.

6. Bruce J. MacFadden and Robert P. Guralnick, "Horses in the Cloud: big data exploration and mining of fossil and extant Equus (Mammalia: Equidae)," *Paleobiology* 43, no. 1 (2016): 1–14.

7. Wendy Williams, *The Horse: The Epic History of Our Noble Companion* (New York: Macmillan, 2015), 1675, Kindle.

8. Green Events, "Whole Earth Man V Horse," accessed Nov 19, 2018. https://www.green-events.co.uk/?mvh_main.

9. Williams, *Horse*, 112–3.

10. Clay McShane and Joel A. Tarr, *The Horse in the City: Living Machines in the Nineteenth Century* (Baltimore: John Hopkins University Press, 2007), 178.

11. Brian Timney and Todd Macuda, "Vision and Hearing in Horses," *Journal of the American Veterinary Medical Association* 218, no. 10 (2001): 1567–74.

12. Williams, *Horse*, 3318–26.

13. Williams, *Horse*, 3318–26.

14. Janet L. Jones, "Becoming a Centaur," *Aeon*, Jan 14, 2022, accessed Mar 7, 2023. https://aeon.co/essays/horse-human-cooperation-is-a-neurobiological-miracle.

15. Janet L. Jones, *Horse Brain, Human Brain: The Neuroscience of Horsemanship* (North Pomfret, VT: Trafalgar Square Books, 2020), 33, Ebook.

16. Jared Diamond, *Guns, Germs and Steel: A Short History of Everybody for the Last 13,000 Years* (New York: Vintage, 1998), 1394, Kindle.

17. David W. Anthony, *The Horse, the Wheel and Language: How Bronze-Age Riders from the Eurasian Steppes Shaped the Modern World* (Princeton, NJ: Princeton University Press, 2007), 303.

18. Anthony, *Horse, Wheel, Language*, 311–2.

19. Lynn Jr., White, *Medieval Technology and Social Change* (London: Oxford University Press, 1962), 27.

20. Lindner, "Chariots" GuiYun Jin et al., "An important military city of the Early Western Zhou Dynasty: Archaeobotanical evidence from the Chenzhuang site, Gaoqing, Shandong Province," *China Science Bulletin* 57, nos. 2–3 (2012): 253–60.

21. Martin Trautmann et al., "First bioanthropological evidence for Yamnaya horsemanship," *Science Advances* 9, no. 9 (Mar 2023).

22. The Yamnaya Impact on Prehistoric Europe Project, accessed July 11, 2023. https://www.helsinki.fi/en/research groups/the-yamnaya-impact-on-prehis toric-europe.

23. Alfred Weber in Karl Jaspers, *The Origin and Goal of History* (New Haven, CT: Yale University Press, 1953), 16. Aruska, "Prehistoric Horses."

24. Íñago Olalde et al., "The genomic history of the Iberian Peninsula over the past 8000 years," *Science* 363 (Mar 2019) (6432): 1230–4.

25. Ben Krause-Kyora et al., "Neolithic and medieval virus genomes reveal complex evolution of hepatitis B," *eLife* 7 (2018).

26. Hodder, *Entangled*, 121. Tim Ingold, *Hunters, Pastoralists and Ranchers: Reindeer Economics and Their Transformations* (Cambridge: Cambridge University Press, 1980), 281.

27. Iosif Lazaridis et al., "A genetic

probe into the ancient and medieval history of Southern Europe and West Asia," *Science*, 377 (2022), 6609: 940–51.

28. Rolf Strootman, "Alexander's Thessalian cavalry," *Talanta* 42/43 (2010–2011), 52.

29. Palaephatues, *On Unbelievable Tales*, trans. Jacob Stern (Wauconda: Bolchazy-Carducci Publishers, 1996), 4.

30. Cierra Tolentino, "Centaurs: Half-Horse Men of Greek Mythology," *History Cooperative*, Oct 17, 2022, accessed Mar 6, 2023. https://historycooperative.org/centaurs/. Some accounts of the Yamnaya invasion have even been described as genocide. See Colin Barras, "Story of the most murderous people of all time revealed in ancient DNA," *New Scientist*, March 27, 2019.

31. Bernal Díaz, *The Conquest of New Spain*, trans. J.M. Cohen (London: Penguin, 1963), 102, Kindle.

32. Olga Gertcyk, "Ancient Mummy with 1100 year old Adidas boots died after she was struck on the head," *The Siberian Times*, Apr 12, 2017.

33. White, *Medieval Technology*, 2.

34. Attributed to B.H. Liddell Hart, in Thomas J. Craughwell, *The Rise and Fall of the Second Largest Empire in History: How Genghis Khan's Mongols Almost Conquered the World* (Beverly, MA: Fair Winds Press, 2010), 152–3.

35. Jack Weatherford, *Genghis Khan and the Making of the Modern World* (New York: Broadway Books, 2005), 183–3, Kindle.

36. Much of the mythology around these invaders comes from the hysterical account of Matthew Paris, a monk who wrote a contemporaneous account. See Weatherford, *Genghis Khan*, 2658–3136.

37. Paul Ratchnevsky, *Genghis Khan: His Life and Legacy*, trans. Thomas Nivison Haining (Oxford: Blackwell, 1991), 185, 122.

38. Weatherford, *Genghis Khan*, 1636.

39. Bat-Ochir Bold, *Mongolian Nomadic Society: A Reconstruction of the "Medieval" History of Mongolia* (New York: St. Martin's Press, 2001), 168n. Glenn Danford Bradley, *The Story of The Pony Express* (Chicago: McClurg & Co., 2010).

40. McLuhan, *Understanding Media*, 99.

41. White, *Medieval Technology*, 11–3.

42. Gilles Deleuze and Félix Guattari, *Nomadology: The War Machine* (Seattle: Wormwood Distribution, 2010), 97.

43. Sun Tzu, *The Art of War*, trans. Lionel Giles (Pax Librorum, 2009), 42.

44. Paolo Virilio, *Speed and Politics: An Essay on Dromology*, trans. M. Polizzotti (South Pasadena, CA: Semiotext(e), 2006), 171.

45. The economist Alfred Weber suggested this. Jaspers, *Origin and Goal*, 16.

46. Lewis Mumford, *The City in History: Its Origins, Its Transformations, and Its Prospects* (New York: Harcourt, Brace & World Inc., 1961), 371.

47. Thorstein Veblen, *The Theory of the Leisure Class* (Oxford: Oxford University Press, 2007), 95–6.

48. Kevin Kelly, *What Technology Wants* (New York: Viking, 2010), 180.

49. Tim Cresswell, *On the Move: Mobility in the Western World* (New York: Routledge, 2006), 59. Rebecca Solnit, *Motion Studies: Time, Space and Eadweard Muybridge* (London: Bloomsbury, 2003), 184–5.

50. McLuhan, *Understanding Media*, 198.

51. Andy Needham et al., "Art by firelight? Using experimental and digital techniques to explore Magdalenian engraved plaquette use at Montastruc (France)," *PloS One* 17, no. 4 (Apr 2022): 1–28.

Chapter 7

1. Dervla Murphy, *Wheels Within Wheels: The Making of a Traveller* (London: Eland Publishing, 2011), 1573–4. Kindle.

2. Louise-Ann Leyland et al., "The effect of cycling on cognitive function and well-being in older adults," *PloS One* 14, no. 2 (2019).

3. Carolos A. Celis-Morales et al., "Association between active commuting and incident cardiovascular disease, cancer, and mortality: prospective cohort study," *BMJ* 357 (Apr 2017): 1456.

4. Edmund King quoted in Carlton Reid, *Roads were not Built for Cars: How Cyclists were the First to Push for Good Roads & Became the Pioneers of Motoring* (Washington, DC: Island Press, 2015), xi.

5. John Berger, *Ways of Seeing* (London: BBC and Penguin Books, 1972), 83–108.

6. Dana Yanocha, "Optimising New Mobility Services," *International Transport Forum Discussion Papers* (Paris: OECD, 2018).

7. *Century Magazine*, XLIX (November 1894–April 1895), 306. *Bicycling World*, XXXII (May 15, 1896), 11. *Detroit Free Press*, quoted in *Literary Digest*, XIII (1896), 197. *Wheeland Cycling TradeReview*, XX (September 17, 1897), 36. See Richard Harmond, "Progress and Flight: An Interpretation of the American Cycle Craze of the 1890s," *Journal of Social History* 5, no. 2 (Winter 1971–1972): 249.

8. Hans-Erhard Lessing, "The evidence against Leonardo's bicycle," *Cycle History* 8 (1998): 49–56.

9. Tony Hadland and Hans-Erhard Lessing, *Bicycle Design: An Illustrated History* (Cambridge, MA: The MIT Press, 2014), 9.

10. Chris Connolly, "How the Bicycle Emancipated Women," *Mental Floss*, Aug 18, 2018, accessed Jun 15, 2023. https://www.mentalfloss.com/article/19373/how-bicycle-emancipated-women.

11. Harmond, "Progress and Flight," 240.

12. Peter Zheutlin, "Women on Wheels: The Bicycle and the Women's Movement of the 1890s," Annielondonderry.com, accessed Jun 18, 2018. annielondonderry.com/womenWheels/html.

13. Glen Norcliffe, *The Ride to Modernity: The Bicycle in Canada, 1869–1900* (Toronto: University of Toronto Press, 2001), 107.

14. Sarah Nipper, "Wheels of Change: The Bicycle and Women's Rights," *MS Magazine*, May 7, 2014.

15. For example, see Maria E Ward, *Bicycling for Ladies* (New York: Bretano's, 1896). Kindle.

16. Quoted in Louise Dawson, "How the Bicycle became a Symbol of Women's Emancipation," *The Guardian*, Nov 4, 2011.

17. McShane and Tarr, *Horse in the City*, 98.

18. Nell Frizell, "I belong on the road as much as any man. Male rage won't scare me off my bike," *The Guardian*, Aug 26, 2015.

19. Connolly, "Emancipated Women."

20. Frizell, "Male rage."

21. Elizabeth Robins Pennell, *Over the Alps on a Bicycle* (Charlotte: Eltanin Publishing, 2014), 827–8, Kindle.

22. Sheila Hanlon, "Imperial Bicyclists: Women travel writers on wheels in the late nineteenth and early twentieth century world," Shelahanlon.com, accessed Oct 13, 2022. http://www.sheilahanlon.com/?p=1343.

23. Peter Zheutlin, *Around the World on Two Wheels: Annie Londonderry's Extraordinary Ride* (New York: Citadel Press, 2007), 231.

24. Peter Zheutlin, "Backstory: Chasing Annie Londonderry," *The Christian Science Monitor*, Aug 28, 2006.

25. Adam Marsal and Brian Fleming, "Riding for your Rights!" *We Love Cycling*, June 10, 2015.

26. Nipper, "Wheels of Change."

27. Frances Willard, *Wheel Within a Wheel: How I Learned to Ride the Bicycle with Some Reflections by the Way* (Bedford, MA: Applewood Books, 1895), 38–9, Kindle.

28. Zheutlin, "Backstory."

29. Giovanni Di Plano Carpini, *The Story of the Mongols Whom we call the Tartars* (Boston: Branden, 1996), 54.

30. Connolly, "Emancipated Women."

31. Marsal and Fleming, "Riding for your Rights!"

32. Joseph Stromberg, "'Bicycle Face': a 19th-century health problem made up to scare women away from biking," *Vox*, Mar 24, 2015, accessed Jun 15, 2023. https://www.vox.com/2014/7/8/5880931/the-19th-century-health-scare-that-told-women-to-worry-about-bicycle.

33. Connolly, "Emancipated Women."

34. Frizell, "Male rage."

35. J. Knowles et al., *Collisions involving pedal cyclists on Britain's roads: establishing the causes* (Wokingham, UK: Transport Research Laboratory, 2009).

36. Rosanna Spero, *RAC Report on Motoring 2012.*

37. For examples see: Michael Hann, "Why I Hate Cyclists," *The Guardian*, Oct 25, 2002. Matthew Parris, "What's Smug and Deserves to be Decapitated?" *The Times*, Dec 27, 2007. Courtland Milloy, "Bicyclist bullies try to rule the road in D. C.," *The Washington Post*, Jul 8, 2014. John Kelly, "Please stop riding your bike

on the sidewalk in D.C.'s central business district," *The Washington Post*, Jul 7, 2014. Mark Hilliard and Áine McMahon, "Ryanair's Michael O'Leary says cyclists should be shot," *The Irish Times*, May 4, 2016.

38. Peter Walker, "Sabotage and hatred: what have people got against cyclists?" *The Guardian*, Jul 1, 2015.

39. Tara Beth Goddard, *Drivers' Attitudes and Behaviors toward Bicyclists: Intermodal Interactions and Implications for Road Safety—Final Report* (Portland, OR: National Institute for Transportation and Communities, 2017), 155.

40. Tom Stafford, "The psychology of why cyclists enrage car drivers," BBC, Feb 12, 2013, accessed Mar 16, 2023. https://www.bbc.com/future/article/20130212-why-you-really-hate-cyclists.

41. According to the U.S. Federal Highway Administration statistics, men drive 61% more miles than women per year. United States Department of Transportation, "Federal Highway Administration," accessed Mar 17, 2023. https://www.fhwa.dot.gov/ohim/onh00/bar8.htm.

42. Ian Walker, "Drivers overtaking bicyclists: Objective data on the effects of riding position, helmet use, vehicle type and apparent gender," *Accident Analysis and Prevention* 39, no. 2 (2007): 417–25.

43. Dawson, "Bicycle Women's Emancipation."

44. Dawson, "Bicycle Women's Emancipation." Terry Slavin, "If there aren't as many women cycling as men...you need better infrastructure," *The Guardian*, Jul 9, 2015.

45. Rachel Alred, "Investigating the Rates and impacts of near misses and related incidents among UK cyclists," *Journal of Transport and Health* 2, no. 3 (Sep 2015): 379–93. Nataly Pinto, *Mujeres en Bici: Una Expersión de Libertad que Transciende Fronteras* (Quito: FES-ILDIS, 2017), 51.

46. Daniel Defraia, "North Korea Bans Women from Riding Bicycles... Again," *cnbc.com*, Jan 17, 2013, accessed June 15, 2023. https://www.cnbc.com/opt-in-check/?pub_referrer=%2Fid%2F100386298.

47. Steven Turner, "Meet Bushra Al-Fusail & the Yemeni women cycling in the face of oppression & civil war," *Huck Magazine*, Jul 26, 2015, accessed Jun 15, 2023. https://www.huckmag.com/article/bushra-al-fusail.

48. Sarah Lazare, "Meet the Yemeni Woman Using Creative Direct Action to Resist the Country's Brutal War," accessed Mar 15, 2016. https://www.alternet.org/world/meet-yemeni-woman-using-creative-direct-action-resist-countrys-brutal-war?fbclid=IwAR2swHuL4bpg1ABkpFdU2_X6alXwqsCpgUxB5p7usuRDiF7oIFqebHTlaUI.

49. Fitz Cahall, "Boundary Breakers Afghan Women's Cycling Team," *National Geographic*, Nov 13, 2015.

50. John Boyle, "Mode Shift: Philadelphia's Two-Wheeled Revolution in Progress," *Bicycle Coalition of Greater Philadelphia*, May 2011.

51. Virilio, *Speed and Politics*, 171.

52. See Mikael Colville-Anderson, *Copenhagenize* (Washington: Island Press, 2018).

53. Sharon Smith, "Engels and the Origin of Women's Oppression," *International Socialist Review* 2, no. 3 (1997).

54. Reid, *Roads*, xv.

Chapter 8

1. Herman Melville, *Moby Dick or The Whale* (New York: Penguin, 2003), 16.

2. Frances Berdan, *The Aztecs of Central Mexico: An Imperial Society* (New York: Rinehart & Winston, 1982), 14.

3. See Thomas Piketty, *Capital in the Twenty-First Century* (Cambridge: Belknap Press, 2014).

4. Wolfgang Schivelbusch, *The Railway Journey: The Industrialization of Time and Space in the Nineteenth Century* (Oakland, CA.: University of California Press, 2014), 90, Kindle.

5. O.S Nock, *World Atlas of Railways* (Bristol: Victoria House Publishing, 1983), 8.

6. Schivelbusch, *Railway Journey*, 829.

7. Schivelbusch, *Railway Journey*, 904.

8. Nail, *Being and Motion*, 25.

9. For example, in the USA in the year 1800 people moved on average 54 yards per day, in the early 2000s, they moved 30 miles. Urry, *Mobilities*, 3–4.

10. Philip Bagwell, *The Transportation Revolution 1770–1985* (London: Routledge, 1974), 125.

11. Mumford, *Technics & Civilization*, 198.

12. Henri Bergson, *Time and Free Will: An Essay on the Immediate Data of Consciousness*, trans. F. L. Pogson (Mineola: Dover, 2014), 49. Italics in original.

13. Bergson, *Metaphysics*, 65.

14. Peter Merriman, *Mobility, Space and Culture* (London: Routledge, 2012), 153.

15. Schivelbusch, *Railway Journey*, 33–44, 428.

16. Urry, *Mobilities*, 92–3.

17. Urry, *Mobilities*, 109.

18. Henri Bergson, *Creative Evolution*, trans. Arthur Mitchell (Digireads, 2011), 153. Gilles Deleuze also writes about cinema presenting us with "duration." Gilles Deleuze, *Cinema 1: The Movement Image* (London and New York: The Athlone Press, 1989), 5.

19. Mark Cousins, *The Story of Film* (London: Pavilion Books, 2011), 13.

20. Quoted in Rudi Volti, *Society and Technological Change* (New York: Worth Publishers, 2014), 22. See also George M. Beard, *American Nervousness: Its Causes and Consequences* (New York: G.P. Putnam's Sons, 1881).

21. Constantine Pecqueur quoted in Schivelbusch, *Railway Journey*, 196.

22. Urry, *Mobilities*, 94.

23. Cresswell, *On the Move*, 85.

24. Marc Augé, *In the Metro* (Minneapolis: University of Minnesota Press, 2002), 9.

25. Schivelbusch, *Railway Journey*, 817–1018.

26. Deleuze and Guattari, *Nomadology*, 5–6.

27. Claude Lévi-Strauss, *The Savage Mind* (London: Weidenfeld and Nicolson, 1966), 78.

28. Lévi-Strauss, *Savage Mind*, 57, 118.

29. Lévi-Strauss, *Savage Mind*, 200–207.

30. Lévi-Strauss, *Savage Mind*, 189.

31. Ovid, *Metamorphoses: A New Verse Translation*, trans. David Raeburn (London: Penguin Classics, 2004), VIII (185), 306, Ebook.

32. McLuhan, *Understanding Media*, 198–9.

33. Chamberlin, *Island*, 17.

34. Urry, *Mobilities*, 150, 136.

35. Urry, *Mobilities*, 147

36. Marc Augé, *Non-Places: Introduction to an Anthropology of Supermodernity*, trans. John Howe (London: Verso, 1995), 94.

37. Stefan Gössling and Andreas Humpe, "The global scale, distribution and growth of aviation: Implications for climate change," *Global Environmental Change* 65 (2020): 9.

Chapter 9

1. McLuhan, *Understanding Media*, 237.

2. Michel de Montaigne, "Our feelings reach out beyond us," in *The Complete Works: Essays, Travel Journal, Letters*, trans by Donald M. Frame (New York: Alfred A. Knopf, 2003), 9.

3. Benjamin Story and Jenna Silber Storey, *Why We Are Restless: On the Modern Quest for Contentment* (Princeton, NJ: Princeton University Press, 2021), 65.

4. Hedges & Company, "How Many Cars are there in the World in 2023?" accessed Mar 5, 2023. https://hedgescompany.com/blog/2021/06/how-many-cars-are-there-in-the-world/.

5. See Diamond, *Guns, Germs, Steel*, 231.

6. Josef Taalbi and Hana Nielsen, "The role of energy infrastructure in shaping early adoption of electric and gasoline cars," *Nat Energy* 6 (2021): 970–6.

7. James Larminie and John Lowry, *Electric Vehicle Technology Explained* (Chichester: John Wiley & Sons Ltd., 2012), 3–4.

8. Steven Parissien, *The Life of the Automobile: A New History of the Motor Car* (London: Atlantic Books, 2013), 393.

9. Daniel Sperling and Deborah Gordon, "Two Billion Cars: Transforming a Culture," *TR News* 259 (Nov/Dec 2008).

10. Parissien, *Life of Automobile*, 14.

11. The Henry Ford Website, accessed Mar 14, 2023. https://www.thehenryford.org/explore/blog/fords-five-dollar-day/.

12. Jonathan Kwitny, "The Great Transportation Conspiracy," *Harper's* 262, no. 1569 (1981): 15.

13. Kwitny, "Great Transportation Conspiracy," 15.

14. Martha J. Bianco, "The Decline of Transit: A Corporate Conspiracy or Failure of Public Policy? The Case of Portland, Oregon," *Journal of Policy History* 9, no. 4 (1997): 450–74.

15. Van Wilkins, "The Conspiracy

Revisited," *The New Electric Railway Journal* (1995): 20.

16. United States District Court, "N.D. Illinois, United States v. National City Lines. 134 F. Supp. 350 (N.D. Ill. 1955)," accessed Apr 4, 2023. https://casetext.com/case/united-states-v-national-city-lines-2.

17. John Howard Kunstler, *The Geography of Nowhere: The Rise and Decline of America's Man-Made Landscape* (New York: Simon & Schuster, 1993), 91–2. Colin Marshall, "Story of Cities #29: Los Angeles and the 'great American streetcar scandal,'" *The Guardian*, Apr 25, 2016.

18. Kwitny, "Great Transportation Conspiracy," 21.

19. James A. Hart, "Du Pont General Motors Case," *Vanderbilt Law Review* 11, no. 389 (1958): 389.

20. McShane and Tar, *Horse in the City*, 172.

21. Parissien, *Life of Automobile*, 192. Nathan Paulus, "Car Ownership Statistics in the U.S.," MoneyGeek, accessed Apr 5, 2023. https://www.moneygeek.com/insurance/auto/car-ownership-statistics/#vehicle-ownership-by-city.

22. Greg Grandin, *Fordlandia: The Rise and Fall of Henry Ford's Forgotten Jungle City* (New York: Metropolitan Books, 2009).

23. Drew Reed, "Fordlandia—The Failure of Henry Ford's Utopian City in the Amazon," *The Guardian*, Aug 19, 2016.

24. Grandin, *Fordlandia*, 322.

25. Parissien, *Life of Automobile*, 75, 89.

26. Michael Dobbs, "Ford and GM Scrutinized for Alleged Nazi Collaboration," *The Washington Post*, Nov 30, 1998.

27. L.J.K. Setright, *Drive On! A Social History of the Motor Car* (London: Granta Books, 2004), 44.

28. Parissien, *Life of Automobile*, 75–6.

29. Parissien, *Life of Automobile*, 75–6.

30. Parissien, *Life of Automobile*, 75.

31. Parissien, *Life of Automobile*, 78.

32. Rudi Volti, *Cars and Culture: The Life Story of a Technology* (Baltimore: Johns Hopkins University Press, 2004), 80.

33. Parissien, *Life of Automobile*, 16.

34. Cresswell, *On the Move*, 86, 94.

35. Frank Bunker Gilbreth and Lillian Moller Gilbreth, *Fatigue Study; the Elimination of Humanity's Greatest Unnecessary Waste, a First Step in Motion Study* (New York: Macmillan, 1919), 181.

36. Leon James and Diane Nahl, *Road Rage and Aggressive Driving: Steering Clear of Highway Warfare* (New York: Prometheus Books, 2000), 52–5.

37. Cresswell, *On the Move*, 261–2.

38. Donald Appleyard, *Livable Streets* (Berkeley, CA: University of California Press, 1981), 15–24.

39. This is the figure for 2015, according to the WHO. World Health Organization, "Road Safety Traffic Deaths," accessed Apr 9, 2023. https://www.who.int/gho/road_safety/mortality/traffic_deaths_number/en/-.

40. Urry, *Mobilities*, 118.

41. Ralph Nader, *Unsafe at Any Speed: The Designed-In Dangers of the American Automobile* (London: Knightsbridge, 1991), 138.

42. Steven E. Landsburg, *The Armchair Economist: Economics and Everyday Life* (London: Simon & Schuster, 2012), 3–12.

43. Sam Peltzman, "The Effects of Automobile Safety Regulation," *Journal of Political Economy* 83, no. 4 (Aug 1975): 677–726.

44. Vanderbilt, *Traffic*, 4899–904.

45. Matthew Raifman and Ernani Choma, "Disparities in Activity and Traffic Fatalities by Race/Ethnicity," *American Journal of Preventive Medicine* 63, no. 2 (2022): 160–7.

46. Laurie Winkless, *Science and the City: The Mechanics Behind the Metropolis* (London: Bloomsbury, 2016), 2885, Kindle.

47. John Bates and David Liebling, *Spaced Out: Perspectives on Parking Policy* (London: RAC Foundation, 2012), vi.

48. Parissien, *Life of Automobile*, 211. Volti, *Cars and Culture*, 52.

49. Richard Sennett, *The Fall of Public Man* (New York: W.W. Norton & Company, 1976), 453, Kindle.

50. Rob Van Haaren, "Assessment of Electric Cars' Range Requirements and Usage Patterns based on Driving Behavior recorded in the National Household Travel Survey of 2009," *Solar Journey USA* (Jul 2012): 27.

51. Yosef Sheffi and Carlos F. Daganzo, "Another 'paradox' of traffic flow," *Transportation Research* 12, no. 1 (1978): 43–6.

52. Laura Cozzi et al., "As their sales continue to rise, SUVs' global CO2 emissions are nearing 1 billion tonnes," *IEA*,

Feb 27, 2023, accessed Feb 22, 2024. https://www.iea.org/commentaries/as-their-sales-continue-to-rise-suvs-global-co2-emissions-are-nearing-1-billion-tonnes.

53. Parissien, *Life of Automobile*, 326.

54. See the Rezvani "Vengance," accessed Apr 8, 2023. https://www.rezvanimotors.com/rezvani-vengeance#vengeance-security-package-military-package.

55. "Automated and Autonomous Driving: Regulation Under Uncertainty," *International Transport Forum* (Paris: OECD, 2015).

Chapter 10

1. James Morris, *Cities* (London: Faber and Faber, 1963), 14.

2. Tim Cresswell, *Place: A Short Introduction* (Malden, MA: Blackwell, 2004), 39.

3. 54% the world's population lives in cities, but 78% in the developed world. Population Reference Bureau, "2016 World Population Fact Sheet," accessed Jun 15, 2023. www.prb.org.

4. Henri Lefebvre, *The Production of Space* (Oxford: Blackwell, 1991), 12.

5. Deleuze and Guattari, *Nomadology*, 50–2.

6. Deleuze and Guattari, *Nomadology*, 97.

7. Rebecca Solnit, *A Field Guide to Getting Lost* (Edinburgh: Canongate, 2017), 89.

8. Kevin Heatherington, "Rhythm and noise: the city, memory and the archive," in *Urban Rhythms: Mobility, Space and Interaction in the Contemporary City*, ed. Robin James Smith and James Heatherington (Oxford: Wiley Blackwell, 2013), 17–33.

9. For example, ancient Benin was designed using fractal geometry about 1200 CE. See Ron Eglash, *African Fractals: Modern Computing and Indigenous Design* (New Brunswick, NJ: Rutgers University Press, 1999), 20–38.

10. Paul Zucker, "Space and Movement in High Baroque City Planning," *Journal of the Society of Architectural Historians* 14, no. 1 (1955): 8–15.

11. Lewis Mumford, *The Culture of Cities* (San Diego: Harcourt Brace Jovanovich Publishers, 1970), 94.

12. Mumford, *Culture of Cities*, 95.

13. See Virilio, *Speed and Politics*.

14. Mumford, *Culture of Cities*, 94.

15. Heinrich Wölfflin, *Renaissance and Baroque* (London: Collins, Fontana Library, 1964), 61, 64.

16. Schivelbusch, *Railway Journey*, 398.

17. Jan Gehl, *Life Between Buildings: Using Public Space* (Washington: Island Press, 2011), 138.

18. See Christopher Alexander et al., *A Pattern Language: Towns, Buildings, Construction* (New York: Oxford University Press, 1978).

19. See Erika Luckert, "Drawings We Have Lived: Mapping Desire Lines in Edmonton," *Constellations* 4, no. 1 (2013): 318–27.

20. Mikhail Chester et al., "Parking infrastructure: energy, emissions, and automobile life-cycle environmental accounting," *Environmental Research Letters* 5, no. 3 (Jul 2010): 034001.

21. John Bates and David Liebling, *Spaced Out: Perspectives on Parking Policy* (London: RAC Foundation, 2012), vi.

22. Jane Jacobs, *The Death and Life of Great American Cities* (New York: Vintage Books, 1961), 347.

23. Donald C. Shoup, "The trouble with minimum parking requirements," *Transportation Research Part A: Policy and Practice* 33, no. 7–8 (1999): 557–8.

24. "The Perilous Politics of Parking," *The Economist*, Apr 6, 2017.

25. Peter Kageyama, *For the Love of Cities: The Love Affair Between People and their Places* (St. Petersburg: Creative Cities Productions, 2011), 38.

26. Augé, *Non-Places*.

27. See Doreen Massey discussed in Merriman, *Mobility, Space, Culture*, 57.

28. Henri Lefebvre, *The Production of Space* (Oxford: Blackwell, 1991), 26.

29. McLuhan, *Understanding Media*, 108.

30. Michael Hermanussen, "Stature of early Europeans," *Hormones: International Journal of Endocrinology and Metabolism* 2, no. 3 (2003):175–8.

31. Edward Relph, *Place and Placelessness* (London: Pion Ltd., 1976), 15.

32. Gaston Bachelard, *The Poetics of Space* (London: Penguin Classics, 2014), 26, Kindle.

33. Bachelard, *Poetics of Space*, 28.

Chapter 11

1. Mumford, *Culture of Cities*, 5.
2. Keith Tester, "Introduction," in *The Flâneur*, ed. Keith Tester (New York: Routledge, 1994), 15–6.
3. "fllâneur," Online Etymological Dictionary, Oct 13, 2017, accessed May 9, 2023. https://www.etymonline.com/word/flaneur.
4. Charles Baudelaire, *The Painter of Modern Life and other Essays*, trans. and ed. Jonathan Mayne (New York: Phaidon Press, 1970), 9.
5. Frédéric Gros, *A Philosophy of Walking* (London: Verso, 2014), 178.
6. See Guy Debord, *Society of the Spectacle* (Detroit: Black & Red, 1970).
7. Debord, *Spectacle*, Paragraph 25, 119.
8. Certeau, *Everyday Life*, 103.
9. Heatherington, "Rhythm and Noise," 17–8.
10. Lauren Elkin, *Flâneuse: Women Walk the City in Paris, New York, Tokyo, Venice and London* (London: Vintage, 2016), 346–9, Kindle.
11. Bellah, Robert N., *Religion in Human Evolution: From the Paleolithic to the Axial Age* (Cambridge: Belknap Press, 2011), 275.
12. Jeremy Roe, *Antoni Gaudí* (New York: Parkstone International, 2019), 98.
13. "Utopia," in *Chambers Dictionary of Etymology*, ed. Robert K. Barnhart (London: Hachette, 2021), 1190.
14. Stephen Greenblatt, *The Swerve: How the World Became Modern* (New York: W.W. Norton & Company, 2011), 227–30.
15. George Rudebusch, "Dramatic Prefiguration in Plato's *Republic*," *Philosophy and Literature* 26, no. 1 (Apr 2002): 75.
16. Merlin Coverley, *Utopia* (Harpenden: Pocket Essentials, 2012), 19.
17. Steven Mithen, *The Prehistory of the Mind: A Search for the Origins of Art, Religion and Science* (London: Thames & Hudson, 1996), 250.
18. For a colorful and exhaustive list of examples see Richard Wilkinson and Kate Pickett, *The Spirit Level: Why Equality is Better for Everyone* (New York: Bloomsbury Press, 2010). Mithen, *Prehistory*, 250.
19. Sahlins, *Stone Age Economics*, 1–40.
20. Deleuze and Guattari, *Nomadology*, 28–9.
21. Katz, *History of Mathematics*, 41.
22. Bellah, *Religion*, 275.
23. S.N. Eisenstadt, "The Axial Age Breakthroughs," in *The Origin and Diversity of Axial Age Civilizations*, ed. S.N. Eisenstadt (New York: State University of New York Press, 1986), 11.
24. See Karen Armstrong, *The Great Transformation: The World in the Time of Buddha, Socrates, Confucius and Jeremiah* (London: Atlantic Books, 2017).
25. Mumford, *City in History*, 203–4.
26. Mumford, *City in History*, 203.
27. Jaspers, *Origin and Goal*. A similar historical view is found in Eric Voegelin's "leap in being." Eric Voeglin, *Order and History* (Columbia, MO: University of Missouri Press, 1901).
28. Jaspers, *Origin and Goal*, 16.
29. Johann P. Arnason, "The Axial Age and its Interpreters: Reopening a Debate," in *Axial Civilizations and World History*, ed. Johann Arnason et al. (Leiden: Brill, 2005), 25–6.
30. Arsuka, "Prehistoric Horses."
31. See Ernst von Lasaulx quoted in Jaspers, *Origin and Goal*, 8.
32. Some writers have attempted to date the beginning of human self-awareness to a certain point in time. See Julian Jaynes, *The Origin of Consciousness in the Breakdown of the Bicameral Mind* (Boston: Mariner, 2000).
33. Mumford, *City in History*, 202.

Chapter 12

1. Aristotle, *Physics* VIII/1/250b15–251a7, in *Complete Works*, 128.
2. Bellah, *Religion*, 333.
3. G.S. Kirk et al., *The Presocratic Philosophers: A Critical History with a Selection of Texts* (Cambridge: Cambridge University Press, 2007), 169.
4. Seneca, *Letters from a Stoic: Epistulae Morales ad Lucilium*, trans. Robin Campbell (London: Penguin Books, 1969), 188.
5. Seneca, *Letters*, 33.
6. Karen Armstrong, *A Short History of Myth* (Edinburgh: Canongate Books, 2005), 100–3. See also Jaynes, *Bicameral Mind*.
7. This is reported by Cicero (T32), in Robin Waterfield, *The First Philosophers:*

The Presocratics and Sophists (Oxford: Oxford University Press, 2000), 10.

8. Kirk et al., *Presocratic Philosophers*, 105–6. W.K.C. Guthrie, *A History of Greek Philosophy* vol. 1 (Cambridge: Cambridge University Press, 1962), 83–6. Armstrong, *Transformation*, 3770–4.

9. Mumford, *City in History*, 203–4.

10. H.D.F. Kitto, *The Greeks* (London: Penguin Books, 1951), 191.

11. Parmenides, Fragment 8, in Guthrie, *History* vol. 2, 26.

12. Bergson, *Time*, 111–4.

13. Heraclitus, Fragment 28, in Kahn, *Art and Thought*, 760.

14. Richard G. Geldard, *Remembering Heraclitus* (London: Lindisfarne Books, 2000), 2.

15. Bellah, *Religion*, 375.

16. Bergson, *Metaphysics*, 65.

17. Eva Brann, *The Logos of Heraclitus* (Philadelphia: Paul Dry Books, Inc., 2011), 5.

18. Sri Aurobindo, *Essays in Philosophy and Yoga, The Complete Works of Sri Aurobindo* vol. 13 (Pondicherry: Sri Aurobindo Ashram Publication Department, 1998), 6085, Kindle.

19. Heraclitus, Fragment 7 in Geldard, *Remembering*, 135.

20. Heraclitus, Fragment 17 in Geldard, *Remembering*, 157.

21. Geldard, *Remembering*, 56.

22. Kahn, *Art and Thought*, 880.

23. Plato, *Symposium* 220d, trans. M.C. Howatson (Cambridge: Cambridge University Press, 2008), 60.

24. Plato, *Apology*, 29d–e, in *Last Days*, 53.

25. *parrhêsia*, *autarkeia*, *karteria*, and *annidein* respectively.

26. Diogenes Laertius, *Lives*, Book VI/32–38, 275–8.

27. Aristotle, *Physics*, VIII/1 250b10–250b14, in *Complete Works*, 128.

28. Aristotle, *Physics*, VIII/1, 251b11–251b28, in *Complete Works*, 130.

29. Aristotle, *Physics*, VIII/1, 252a6–252b6, in *Complete Works*, 131.

30. Diogenes Laertius, *Lives* Book X/136–7, 536–7.

31. Rovelli, *Reality*, 1846–8.

32. Lucretius, *On the Nature of Things*, trans. William Ellery Leonard (Global Grey, 2018), 145, Kindle.

33. Lucretius, *Nature of Things*, 360.

34. Lucretius, *Nature of Things*, 394.

35. Thomas Nail, *Lucretius I: An Ontology of Motion* (Edinburgh: Edinburgh University Press, 2018), 163–5.

36. Rovelli, *Reality*, 222.

37. Rovelli, *Reality*, 207–10.

38. Nail, *Lucretius*, 62.

Chapter 13

1. Anonymous from the *Ts'ai-ken tan*, quoted in Fritjof Capra, *The Tao of Physics: An Exploration of the Parallels Between Modern Physics and Eastern Mysticism* (Boulder, CO: Shambhala, 1975), 194.

2. Armstrong, *Transformation*, 1345–59.

3. Martin Kern, "Fenshu kengru," in *The Encyclopedia of Confucianism*, ed. Yao Xinzhong (New York: Routledge, 2003), 213–4.

4. Ronnie L. Littlejohn, *Confucianism: An Introduction* (London: I.B. Tauris, 2011), 27.

5. Mo Zi, *The Book of Master Mo*, trans. Ian Johnston (London: Penguin Classics, 2013), 752, Kindle.

6. *Mengzi* 2A6, quoted in Littlejohn, *Confucianism*, 53.

7. Confucius, *The Analects of Confucius* 12.2, trans. Burton Watson (New York: Columbia University Press, 2007), 80.

8. Simon Blackburn, *Ethics: A Very Short Introduction* (Oxford: Oxford University Press, 2001), 101.

9. Confucius, *Analects* 15.5, 106.

10. Quoted in Littlejohn, *Confucianism*, 95.

11. Chuang Tzu, *Chuang Tzu: The Inner Chapters*, trans. A.C. Graham (Indianapolis: Hackett Publishing Company, 1989), 89.

12. Lao Tzu, *Tao Te Ching: The Book of the Way*, trans. Sam Torode (Nashville: Ancient Renewal, 2018), 40.

13. Lao Tzu, *Tao Te Ching*, 1.

14. Plato, *Apology*, 21d in *Last Days*, 42.

15. *Xunzi*, 21, quoted in Bellah, *Religion*, 468–9.

16. Chung-ying Cheng, "Qi (Ch'i): Vital Force," in *Encyclopedia of Chinese Philosophy*, ed. Antonio S. Cua (New York: Routledge, 2003), 616.

17. Littlejohn, *Confucianism*, xv.

18. Michio Kushi and Phillip Jannetta, *Macrobiótica y Medicina Oriental*

(Maldonado, Uruguay: Publicaciones GEA, 1992), 29.

19. David Shepherd Nivison, "Nourishment of *Qi* and Ethical Values," in Michael Loewe and Edward L. Shaughnessy eds., *The Cambridge History of Ancient China: From the Origins of Civilization to 221 BC* (Cambridge: Cambridge University Press, 1999), 775–6.

20. Wyatt, Don J. "Guan wu (observation of things)," in Xinzhong, *Encyclopedia of Confucianism*, 233–4.

21. Jordan Jacobs, *Taoism: A Friendly Beginners Guide On Taoism And Taoist Beliefs* (Relentless Progress Publishing, 2015), 410–19, Kindle.

22. Lars Svendsen, *A Philosophy of Boredom* (London: Reaktion Books, 2005), 49–106.

23. Siegfried Kracauer quoted in Kate Greene, "Planet Boredom," *Aeon*, Feb 26, 2014, accessed May 17, 2023. https://aeon.co/essays/what-four-months-on-mars-taught-me-about-boredom.

24. Quoted in Jacobs, *Taoism*, 268–9.

25. Capra, *Tao of Physics*, 191.

26. Littlejohn, *Confucianism*, 9.

27. Will Buckingham, "The Uncertainty Machine," *Aeon*, Oct 11, 2013, accessed May 15, 2023. https://aeon.co/essays/forget-prophecy-the-i-ching-is-an-uncertainty-machine.

28. Capra, *Tao of Physics*, 3762–72.

29. Richard Wilhelm in *I Ching or Book of Changes*, trans. Richard Wilhelm (London: Arkana, 1989), 1.

30. *I Ching*, 280–1.

31. Lau Tzu, *Tao Te Ching*, 14.

32. Heraclitus, Fragment 7, in Geldard, *Remembering*, 135.

33. Arnold, J. Toynbee, *A Study of History*, abridged ed. (Oxford: Oxford University Press, 1987), 51.

34. Capra, *Tao of Physics*, 24.

35. Carl Gustav Jung in *I Ching*, xxiv.

36. Rovelli, *Reality*, 2752–3.

37. John A. Wheeler, "Physics in Copenhagen in 1934 and 1935," in *Niels Bohr: a centenary volume*, ed. A.P. French (Cambridge, MA: Harvard University Press, 1985), 224.

38. Chuang Tzu, *Book of Chuang Tzu*, 2244.

39. Chuang Tzu, *Book of Chuang Tzu*, 2219–21.

40. Lao Tzu, *Tao Te Ching*, 40.

41. Chuang Tzu, *Book of Chuang Tzu*, 795–8.

42. Chuang Tzu, *Book of Chuang Tzu*, 2320–6.

43. David Shepherd Nivison, "'Zheng Ming,'" in Loewe and Shaughnessy eds, *Cambridge History of Ancient China*, 797.

44. Lao Tzu, *Tao Te Ching: The Classic Book of Integrity and the Way* 30 (80), trans. Victor H. Mair (New York: Bantam Books, 1990), 79.

Chapter 14

1. Addy Pross, "Life's restlessness," *Aeon*, Apr 29, 2014, accessed May 19, 2023. https://aeon.co/essays/paradoxes-of-stability-how-life-began-and-why-it-can-t-rest.

2. A.R.P. Rau, "The Rosetta-Philae Comet Mission as Physics Appreciation," *Resonance* 20 (4) (Apr 2014): 346–51.

3. For more see Katie Mack, *The End of Everything (Astrophysically Speaking)* (New York: Simon & Schuster, Inc., 2020), 1260–363, Kindle.

4. Huw Price, *Time's Arrow and Archimedes' Point: New Directions in the Physics of Time* (Oxford: Oxford University Press, 1996), 7, 37.

5. See for example Richard Dawkins, *The Selfish Gene* (Oxford: Oxford University Press, 1976).

6. Iain McGilchrist, *The Master and his Emissary: The Divided Brain and the Making of the Western World* (New Haven and London: Yale University Press, 2018), 128.

7. Capra, *Tao of Physics*, 2621–7.

8. Capra, *Tao of Physics*, 3265–8.

9. Walter Evans-Wentz quoted in Peter Matthiessen, *The Snow Leopard* (London: Vintage, 2003), 67.

10. Plutarch: *Life of Alexander*, 65.I, quoted in *The Cynic Philosophers from Diogenes to Julian*, ed. Robert Dobbin (London: Penguin, 2012), 247–8.

11. Toynbee, *Study of History*, 21.

12. Vishvapani Blomfield, *Gautama Buddha: The Life and Teachings of the Awakened One* (London: Quercus, 2011), 94–100.

13. *Angulimāla Sutta* in *The Middle Length Discourses of the Buddha: A Translation of the Majjhima Nikāya*, trans. Bhikkhu Ñanamole and Bhikkhu Bodhi (Boston: Wisdom Publications, 1995), 710–17.

14. Sangharakshita, *Who is the Buddha?* in *Complete Works* vol. 3 (Cambridge: Windhorse Publications, 2017), 148.

15. Amira Mittermaier, *Dreams That Matter: Egyptian Landscapes of the Imagination* (Berkeley, CA: University of California Press, 2011), 128.

16. Samuel Beckett, "A Piece of Monologue," in *The Complete Dramatic Works* (London: Faber and Faber Ltd., 1986), 494.

17. Sometimes referred to as "conditioned co-production." The Sanskrit term is *pratītyasamutpāda*.

18. Edward Conze, *Buddhism: Its Essence and Development* (Birmingham: Windhorse, 2001), 33.

19. Hodder, *Entangled*, 169.

20. Sangharakshita, *A Survey of Buddhism: Its Doctrines and Methods Through the Ages* (Birmingham: Windhorse, 2001), 77–9.

21. *The Holy Teaching of Vimalakirti: A Mahayana Scripture*, trans. Robert A.F. Thuman (Philadelphia: Pennsylvania State University Press, 1976), 56.

22. These are the *Hinayana* (lesser vehicle), *Mahayana* (greater vehicle), and *Vajrayana* (diamond vehicle). See Andrew Skilton, *A Concise History of Buddhism* (Cambridge: Windhorse Publications Ltd., 2013), 80–98, Epub.

23. Werner Heisenberg, *Encounters with Einstein And Other Essays on People, Places, and Particles* (Princeton, NJ: Princeton University Press, 1983), 53.

24. Jim Al-Khalili, *Quantum: A Guide for the Perplexed* (London: Weidenfeld & Nicolson, 2003), 16–20.

25. Quoted in Sangharakshita, *The Bodhisattva Ideal* (Cambridge: Windhorse, 1999), 214.

26. Urry, *Mobilities*, 38.

27. Carl Gustav Jung, *Modern Man in Search of a Soul* (New York: Harvest, 1933), 74.

28. Henri Bergson, *Matter and Memory*, trans. Nancy Margaret Paul and W. Scott Palmer (New York: Zone Books, 1991), 197.

Chapter 15

1. Edmund Husserl, "World of the Living Present," in *Shorter Works* (Notre Dame, IN: UNDP, 1981), 248.

2. See Rovelli, *Reality*, 2715–2801.

3. *Mahāparinibbāna Sutta* DN16.6.7., in *The Long Discourses of the Buddha: A Translation of the Dīgha Nikāya*, trans. Maurice Walshe (Boston: Wisdom Publications, 1995), 270.

4. *The Dhammapada: The Way of Truth*, trans. Sangharakshita (Cambridge: Windhorse, 2001), 13.

5. See Leigh Brasington, *Right Concentration: A Practical Guide to the Jhānas* (Boulder, CO: Shambhala, 2015).

6. Oppenheimer, *Out of Eden*, 791.

7. Mithen, *Prehistory*, 248.

8. Maurice Merleau-Ponty, *Phenomenology of Perception*, trans. Colin Smith (London: Routledge, 2002), 240–82.

9. John R. Searle, *Intentionality: An Essay in the Philosophy of Mind* (Cambridge: Cambridge University Press, 1983), 84.

10. Merleau-Ponty, *Phenomenology of Perception*, 94–5.

11. Armstrong, *Transformation*, 2541.

12. Heraclitus, Fragment 20, in Geldard, *Remembering*, 157.

13. Welter, Albert. "Mahākāśyapa's Smile: Silent Transmission and the Kung-an (Kōan) Tradition," in *The Kōan: Texts and Contexts in Zen Buddhism*, ed. Steven Heine and Dale S. Wright (New York: Oxford University Press, 2000), 77.

14. Case 43, *The Gateless Gate: The Classic Book of Zen Koans*, trans. Kōun Yamada (Boston: Wisdom Publications, 2004), 477.

15. Heraclitus, Fragment 49a, in Guthrie, *History*, vol. 2, 24.

16. Massumi, *Parables*, 10.

17. Massumi, *Parables*, 201.

18. Anil Ananthaswamy, "Time flows uphill for remote Papua New Guinea tribe," *New Scientist*, May 30, 2012. Rafael E. Núñez and Eve Sweetser, "With the Future Behind Them: Convergent Evidence From Aymara Language and Gesture in the Crosslinguistic Comparison of Spatial Construals of Time," *Cognitive Science* 30, no. 3 (2006): 401–50.

19. I have chosen *advaita* rather than *advaya* here, the latter simply meaning "not-two." *Advaita*, refers more to an ontological non-duality rather than the epistemic meaning. See T.R.V. Murti, *The Central Philosophy of Buddhism* (London: Routledge, 2008), 646–50.

20. W.K.C. Guthrie, *The Greek Philosophers: from Thales to Aristotle* (London: Routledge, 2013), 44–5.
21. Lao Tzu, *Tao Te Ching*, 1. Sangharakshita, *Survey*, 85–6.
22. Bruno Losch, "Family Farming: At the Core of World's Agricultural History," in *Family Farming and the Worlds to Come*, ed. Jean-Michel Sourisseau (Paris: Springer, 2015), 13–36.
23. Mumford, *City in History*, 202–3.
24. Bill Hatcher, "Why You Move the Way You Do," *National Geographic*, Feb 28, 2016.
25. Oswald Spengler, *The Decline of the West* vol. 2 (Petosky, MI: Random Shack, 2016), 9930, Kindle. Emphasis in original.
26. Robert E. and Donald S. Lopez, Jr., "tathatā," in *The Princeton Dictionary of Buddhism* (Princeton, NJ: Princeton University Press, 2014), 899.
27. Thomas Hobbes, *Leviathan* (Oxford: Oxford University Press, 1966), 39.
28. Hobbes, *Leviathan*, 129, 139.
29. Maria-Gabriele Wosien, *Sacred Dance: Encounter with the Gods* (New York: Thames and Hudson, 1986), 11, 21.
30. From the Greek *ekstasis*.
31. Rudolph von Laban quoted in Evelyn Dörr and Lori Lantz, "Rudolf von Laban: The 'Founding Father' of Expressionist Dance," in *Dance Chronicle* 26, no. 1 (2003): 9.
32. Giovanni-Andrea Gallini, *Critical Observations on the Art of Dancing* (London: 1765), 4, Kindle.
33. Gallini, *Observations*, 8–9.
34. Jane Goodall, "Primate spirituality," in *Encyclopedia of Religion and Nature*, ed. Bron Taylor (New York: Continuum, 2005), 1303–6.
35. Sachs, *Dance*, 3.
36. W.B. Yeats, "Among School Children," in *The Collected Poems of W. B. Yeats*, ed. Richard J. Finneran (New York: Palgrave Macmillan, 1989), 217.

Bibliography

Adams, Robert McCormick. *Heartland of Cities: Surveys of Ancient Settlement and Land Use on the Central Floodplain of the Euphrates.* Chicago: University of Chicago Press, 1981.

Alentorn-Geli, Eduard, et al. "The Association of Recreational and Competitive Running With Hip and Knee Osteoarthritis: A Systematic Review and Meta-analysis." *Journal of Orthopaedic & Sports Physical Therapy* 47 no. 6 (June 2017): 373–90.

Alexander, Christopher, et al. *A Pattern Language: Towns, Buildings, Construction.* New York: Oxford University Press, 1978.

Al-Khalili, Jim. *Quantum: A Guide for the Perplexed.* London: Weidenfeld & Nicolson, 2003.

Alred, Rachel. "Investigating the Rates and impacts of near misses and related incidents among UK cyclists." *Journal of Transport and Health* 2, no. 3 (Sep 2015): 379–93.

Altares, Guillermo. "Una humanidad vagabunda: en el futuro, todos volveremos a ser nómadas." *El País,* Nov 24, 2022.

Amato, Joseph. *On Foot: A History of Walking.* New York: New York University Press, 2004.

Ananthaswamy, Anil. "Time flows uphill for remote Papua New Guinea tribe." *New Scientist,* May 30, 2012.

Andersen, Jen Jacobs. "Research: Women are Better Runners than Men." Run Repeat. Accessed 6 Feb, 2023. https://runrepeat.com/research-women-are-better-runners-than-men.

The Andrew W. Mellon Foundation. "In South Africa, Discovering the World's Oldest Drawing." December 2018. Accessed Jan 27, 2023. https://web.archive.org/web/20200309004930/https://mellon.org/shared-experiences-blog/archaeologists-discovery-worlds-oldest-drawing-highlights-strong-interest-ancient-rock-art/#top.

Anthony, David W. *The Horse, the Wheel and Language: How Bronze-Age Riders from the Eurasian Steppes Shaped the Modern World.* Princeton: Princeton University Press, 2007.

Appleyard, Donald. *Livable Streets.* Berkeley, CA: University of California Press, 1981.

Aristotle. *Aristotle: The Complete Works of Aristotle* (The Revised Oxford Translation). Translated by Jonathan Barnes. Oxford: Oxford University Press, 1991.

Armstrong, Karen. *The Great Transformation: The World in the Time of Buddha, Socrates, Confucius and Jeremiah.* London: Atlantic Books, 2017. Kindle.

______. *A Short History of Myth.* Edinburgh: Canongate Books, 2005.

Arnason, Johann, et al., eds. *Axial Civilizations and World History.* Leiden: Brill, 2005.

Arsuka, Nirwan Ahmad. "A Tale of Prehistoric Horses in South Sulawesi." *The Jakarta Post,* Dec 8, 2015.

Artress, Lauren. *Walking a Sacred Path: Rediscovering the Labyrinth as a Spiritual Practice.* New York: Riverhead Books, 2006.

Augé, Marc. *In the Metro.* Minneapolis: University of Minnesota Press, 2002.

______. *Non-Places: Introduction to an Anthropology of Supermodernity.* Translated by John Howe. London: Verso, 1995.

Azéma, Marc, and Florent Rivère. "Animation in Paleolithic Art: A Pre-Echo of Cinema." *Antiquity* 86 (2012): 316–24.

Bachelard, Gaston. *The Poetics of Space*. London: Penguin Classics, 2014. Kindle.
Bagwell, Philip. *The Transportation Revolution 1770–1985*. London: Routledge, 1974.
Balter, Michael. "The Tangled Roots of Agriculture." *Science* 327, no. 5964 (2010): 404–6.
______. "Why Settle Down? The Mystery of Communities." *Science* 282, no. 5393 (Nov 1998): 1442–5.
Bannister, Roger. *The First Four Minutes*. Stroud: Sutton Publishing Ltd., 2004.
Bardi, Ugo, et al. "Toward a General Theory of Societal Collapse: A Biophysical Examination of Tainter's Model of the Diminishing Returns of Complexity." *BioPhysical Economics and Resource Quality* 4, no. 3 (2019).
Barker, Graeme. *The Agricultural Revolution in Prehistory: Why Did Foragers Become Farmers?* Oxford: Oxford University Press, 2006.
Barnhart, Robert K., ed. *Chambers Dictionary of Etymology*. London: Hachette, 2021.
Barnish, Maxwell S., and Jean Barnish. "High-heeled shoes and musculoskeletal injuries: a narrative systematic review." *BMJ Open* 6, no. 1 (Jan 2016): 1–8.
Barras, Colin. "Story of the most murderous people of all time revealed in ancient DNA." *New Scientist,* March 27, 2019.
Barrera Loredo, Jacob, et al. "Influence of High Heels on Walking Motion: Gait Analysis." *Journal of Applied Biomechanics,* Dec 2015.
Bar-Yosef, Ofer. "Climatic Fluctuations and Early Farming in West and East Asia." *Current Anthropology* 52, no. S4 (2011): S175–93.
Bates, John, and David Liebling. *Spaced Out: Perspectives on Parking Policy*. London: RAC Foundation, 2012.
Batuman, Elif. "The Sanctuary: The World's Oldest Temple and the Dawn of Civilization." *The New Yorker,* Dec 19, 2011.
Baudelaire, Charles. *The Painter of Modern Life and other Essays*. Translated and edited by Jonathan Mayne. New York: Phaidon Press, 1970.
Bayne, Tim. *Thought: A Very Short Introduction*. Oxford: Oxford University Press, 2013.
Beard, George M. *American Nervousness: Its Causes and Consequences*. New York: G.P. Putnam's Sons, 1881.
Beckett, Samuel. *The Complete Dramatic Works*. London: Faber & Faber Ltd., 1986.
Beevers, Robert. *The Garden City Utopia*. London: Macmillan, 1988.
Bellah, Robert N. *Religion in Human Evolution: From the Paleolithic to the Axial Age*. Cambridge: Belknap Press, 2011.
Bellah, Robert N., and Hans Joas, eds. *The Axial Age and Its Consequences*. Cambridge: Belknap Press, 2012.
Bellwood, Peter. *First Farmers. The Origins of Agricultural Societies*. Malden: Blackwell Publishing, 2005.
______. *First Migrants: Ancient Migration in Global Perspective*. Chichester: John Wiley & Sons, 2013.
Benedict, Ruth. *Patterns of Culture*. New York: Mentor, 1960.
Benjamin, Walter. *The Arcades Project*. Cambridge: Belknap Press, 1999.
Berdan, Frances. *The Aztecs of Central Mexico: An Imperial Society*. New York: Rinehart & Winston, 1982.
Berger, John. *Ways of Seeing*. London: BBC and Penguin Books, 1972.
Bergson, Henri. *Creative Evolution*. Translated by Arthur Mitchell. Digireads, 2011.
______. *The Creative Mind: An Introduction to Metaphysics*. Translated by T.E. Hulme. New York: Putnam's Sons, 1912. Kindle.
______. *Matter and Memory*. Translated by Nancy Margaret Paul and W. Scott Palmer. New York: Zone Books, 1991.
______. *Time and Free Will: An Essay on the Immediate Data of Consciousness*. Translated by F.L. Pogson. Mineola, TX: Dover, 2014.
Berman, Marshall. *All That Is Solid Melts into Air: The Experience of Modernity*. New York: Penguin Books, 1988.
Bianco, Martha J. "The Decline of Transit: A Corporate Conspiracy or Failure of Public Policy? The Case of Portland, Oregon." *Journal of Policy History* 9, no. 4 (1997): 450–74.
Black, Jeremy. *Visions of the World: A History of Maps*. London: Mitchell Beazley, 2003.

Blackburn, Simon. *Ethics: A Very Short Introduction*. Oxford: Oxford University Press, 2001.
Blomfield, Vishvapani. *Gautama Buddha: The Life and Teachings of the Awakened One*. London: Quercus, 2011.
Bold, Bat-Ochir. *Mongolian Nomadic Society: A Reconstruction of the "Medieval" History of Mongolia*. New York: St. Martin's Press, 2001.
Bond, Mary. *The New Rules of Posture: How to Sit, Stand, and Move in the Modern World*. Rochester, VT: Healing Arts Press, 2007.
Boyle, John. *Mode Shift: Philadelphia's Two-Wheeled Revolution in Progress*. Philadelphia: Bicycle Coalition of Greater Philadelphia, 2011.
Bradley, Glenn Danford. *The Story of The Pony Express*. Chicago: McClurg & Co., 2010.
Bramble, Dennis, and Daniel Lieberman. "Endurance Running and the Evolution of *Homo*." *Nature* 432, no. 7015 (Nov 2004): 345–52.
Bramble, D.M., and F.A. Jenkins, Jr. "Mammalian Locomotor-Respiratory Integration: Implications for Diaphragmatic and Pulmonary Design." *Science* 262, no. 5131 (Oct 1993): 196–7.
Brandt, Jobst. *The Bicycle Wheel*. Menlo Park, CA: Avocet Inc., 1983.
Brann, Eva. *The Logos of Heraclitus*. Philadelphia: Paul Dry Books, Inc., 2011.
Brasington, Leigh. *Right Concentration: A Practical Guide to the Jhānas*. Boulder, CO: Shambhala, 2015.
Bronowski, Jacob. *The Ascent of Man*. London: BBC Books, 1973.
Buckingham, Will. "The Uncertainty Machine." *Aeon*, Oct 11, 2013. Accessed May 15, 2023. https://aeon.co/essays/forget-prophecy-the-i-ching-is-an-uncertainty-machine.
Buswell, Robert E., and Donald S. Lopez, Jr. *The Princeton Dictionary of Buddhism*. Princeton, NJ: Princeton University Press, 2014.
Cahall, Fitz. "Boundary Breakers Afghan Women's Cycling Team." *National Geographic*, Nov 13, 2015.
Callen, Kenneth E. "Mental and emotional aspects of long-distance running." *Psychosomatics* 24, no. 2 (1983): 133–51.
Campbell, I.C. (1995). "The Lateen Sail in World History." *Journal of World History* 6 (1): 1–23.
Cann, Rebecca, et al. "Mitochondrial DNA and Human Evolution." *Nature* 325 (1987): 31–6.
Capra, Fritjof. *The Tao of Physics: An Exploration of the Parallels Between Modern Physics and Eastern Mysticism*. Boulder, CO: Shambhala, 1975. Kindle.
Cauvin, Jacques. *The Birth of the Gods and the Origin of Agriculture*. Cambridge: Cambridge University Press, 2002.
Celis-Morales, Carlos A., et al. "Association between active commuting and incident cardiovascular disease, cancer, and mortality: prospective cohort study." *BMJ* 357 (April 19, 2017): 1456.
Certeau, Michel de. *The Practice of Everyday Life*. Berkeley: University of California Press, 1984.
Chakraborty, Rabindra N. "Sharing Culture and Resource Conservation in Hunter-Gatherer Societies." *Oxford Economic Papers* 59, no. 1 (Jan 2007): 63–88.
Chamberlin, J. Edward. *Horse: How the Horse Has Shaped Civilisations*. Oxford: Signal Books, 2007.
______. *Island: How Islands Transform the World*. London: Elliott & Thompson, 2013.
Chan, Eva K.F., et al. "Human origins in a southern African palaeo-wetland and first migrations." *Nature* 575 (Nov 2019): 185–201.
Chardin, Pierre Teilhard de. *The Phenomenon of Man*. New York: Harper Perennial Modern Classics, 2008.
Chatwin, Bruce. *The Songlines*. New York: Penguin, 1987.
Chester, Mikhail, et al. "Parking infrastructure: energy, emissions, and automobile life-cycle environmental accounting." *Environmental Research Letters* 5, no. 3 (July 2010): 034001.
Chuang Tzu. *The Book of Chuang Tzu*. Translated by Martin Palmer. London: Penguin Classics, 2006. Kindle.

______. *Chuang Tzu: The Inner Chapters*. Translated by A.C. Graham. Indianapolis: Hackett Publishing Company, 1989.

Ciochon, Russell L., and O. Frank Huffman. "Java Man." In *Encyclopedia of Global Archaeology*, edited by Claire Smith, 4182–8. New York: Springer International Publishing, 2020.

Cioran, E.M. *History and Utopia*. Translated by Richard Howard. New York: Arcade Publishing, 2015.

Colville-Anderson, Mikael. *Copenhagenize*. Washington, D.C.: Island Press, 2018.

Confucius. *The Analects of Confucius*. Translated by Burton Watson. New York: Columbia University Press, 2007.

Connolly, Chris. "How the Bicycle Emancipated Women." *Mental Floss*, Aug 18, 2018. Accessed June 15, 2023. https://www.mentalfloss.com/article/19373/how-bicycle-emancipated-women.

Conze, Edward. *Buddhism: Its Essence and Development*. Birmingham: Windhorse, 2001.

Cousins, Mark. *The Story of Film*. London: Pavilion Books, 2011.

Coverley, Merlin. *Psychogeography*. Harpenden: Pocket Essentials, 2006.

______. *Utopia*. Harpenden: Pocket Essentials, 2012.

Craughwell, Thomas J. *The Rise and Fall of the Second Largest Empire in History: How Genghis Khan's Mongols Almost Conquered the World*. Beverly, MA: Fair Winds Press, 2010.

Crawford, Harriet, ed. *The Sumerian World*. Oxon, UK: Routledge, 2013.

Cresswell, Tim. *On the Move: Mobility in the Western World*. New York: Routledge, 2006.

______. *Place: A Short Introduction*. Malden, MA: Blackwell, 2004.

Csikszentmihalyi, Mihaly. *Flow: The Psychology of Optimal Experience*. New York: Harper Perennial, 2008.

Cua, Antonio S., ed. *Encyclopedia of Chinese Philosophy*. New York: Routledge, 2003.

Curry, Andrew. "Göbekli Tepe: The World's First Temple?" *Smithsonian Magazine*, Nov. 2008. Accessed Feb 23, 2023. https://www.smithsonianmag.com/history/gobekli-tepe-the-worlds-first-temple-83613665/.

Darwin, Charles. *The Descent of Man, and Selection in Relation to Sex*. London: John Murray, 1871. Kindle.

______. *On the Origin of Species by Means of Natural Selection; or, the Preservation of Favoured Races in the Struggle for Life*, 2nd ed. Hardbooks, 2016. Kindle.

Davis, Matthew. "Can inequality be blamed on the Agricultural Revolution?" *World Economic Forum*, Oct 25, 2018. Accessed Feb 20, 2023. https://www.weforum.org/agenda/2018/10/how-the-agricultural-revolution-made-us-inequal.

Dawkins, Richard. *The Selfish Gene*. Oxford: Oxford University Press, 1976.

Dawson, Louise. "How the Bicycle Became a Symbol of Women's Emancipation." *The Guardian*, Nov 4, 2011.

Debord, Guy. *Society of the Spectacle*. Detroit: Black & Red, 1970.

______. *Theory of the Dérive*. Situationist International Anthology, 2006.

Defraia, Daniel. "North Korea Bans Women from Riding Bicycles... Again." *cnbc.com*, Jan 17, 2013. Accessed Jun 15, 2023. https://www.cnbc.com/opt-in-check/?pub_referrer=%2Fid%2F100386298.

Deleuze, Gilles. *Cinema 1: The Movement Image*. London: The Athlone Press, 1989.

Deleuze, Gilles, and Félix Guattari. *Nomadology: The War Machine*. Seattle: Wormwood Distribution, 2010.

Demoule, Jean-Paul. *Homo Migrans: De la Sortie d'Afrique au Grand Confinemont*. Paris: Payot, 2022.

D'Errico, Francesco, et al. "Early evidence of San material culture represented by organic artifacts from Border Cave, South Africa." *PNAS* 109, no. 33 (2012): 13214–9.

The Dhammapada: The Way of Truth. Translated by Sangharakshita. Cambridge: Windhorse, 2001.

Diamond, Jared. *Guns, Germs and Steel: A Short History of Everybody for the Last 13,000 Years*. New York: Vintage, 1998. Kindle.

______. "The Worst Mistake in the History of the Human Race." *Discover Magazine*, May 1, 1999. Accessed Feb 20, 2023. https://www.discovermagazine.com/planet-earth/the-worst-mistake-in-the-history-of-the-human-race.

Díaz, Bernal. *The Conquest of New Spain*. Translated by J.M. Cohen. London: Penguin, 1963. Kindle.
Dillon, Sally L., et al. "Domestication to Crop Improvement: Genetic Resources for *Sorghum* and *Saccharum* (Andropogoneae)." *Annals of Botany* 100, no. 5 (Oct 2007): 975–89.
Diogenes Laertius. *Lives of the Eminent Philosophers*. Translated by Pamela Mensch. Oxford: Oxford University Press, 2018.
Di Plano Carpini, Giovanni. *The Story of the Mongols Whom we call the Tartars*. Boston: Branden, 1996.
Dobbin, Robert, ed. *The Cynic Philosophers from Diogenes to Julian*. London: Penguin, 2012.
Dobbs, Michael. "Ford and GM Scrutinized for Alleged Nazi Collaboration." *The Washington Post*, Nov 30, 1998.
Doob, Penelope Reed. *The Idea of the Labyrinth: From Classical Antiquity to the Middle Ages*. Ithaca, NY: Cornell University Press, 1992.
Dörr, Evelyn, and Lori Lantz. "Rudolf von Laban: The 'Founding Father' of Expressionist Dance." In *Dance Chronicle* 26, no. 1 (2003): 1–29.
Draper, Patricia. "!Kung Women: Contrasts in Sexual Egalitarianism in Foraging and Sedentary Contexts." In *Toward an Anthropology of Women*, 77–109. Edited by R.R. Reiter. New York: Monthly Review Press, 1975.
Earls, James. *Born to Walk: Myofascial Efficiency and the Body in Movement*. Chichester: Lotus Publishing, 2014. Kindle.
Eglash, Ron. *African Fractals: Modern Computing and Indigenous Design*. New Brunswick, NJ: Rutgers University Press, 1999.
Eisenstadt, S. N., ed. *The Origin and Diversity of Axial Age Civilizations*. New York: State University of New York Press, 1986.
Eliot, T.S. *Four Quartets*. New York: Harvest, 1943.
Elkin, Lauren. *Flâneuse: Women Walk the City in Paris, New York, Tokyo, Venice and London*. London: Vintage, 2016.
Erlandson, Jon M., and Scott M. Fitzpatrick. "Oceans, Islands, and Coasts: Current Perspectives on the Role of the Sea in Human Prehistory." *Journal of Island & Coastal Archaeology* 1, no. 1 (Feb 2006): 5–32.
Felton, Silke, and Heike Becke. *A Gender Perspective on the Status of the San in Southern Africa*. Windhoek: Legal Assistance Centre, 2001.
Foucault, Michel. "Of Other Spaces, Heterotopias." *Architecture/Mouvement/Continuiteì* 5 (1994): 46–9.
French A.P., ed. *Niels Bohr: A Centenary Volume*. Cambridge, MA: Harvard University Press, 1985.
Frisby, David, and Mike Featherstone, eds. *Simmel on Culture: Selected Writings*. New York: Sage Publications, 1997.
Frizell, Nell. "I belong on the road as much as any man. Male rage won't scare me off my bike." *The Guardian*, Aug 26, 2015.
Gallini, Giovanni-Andrea. *Critical Observations on the Art of Dancing*. London: 1765; Hardpress, 2018. Kindle.
Gamble, Clive. *Timewalkers: The Prehistory of Global Colonization*. Phoenix: Alan Sutton Publishing, 1993.
The Gateless Gate: The Classic Book of Zen Koans. Translated with Commentary by Kōun Yamada. Boston: Wisdom Publications, 2004.
Gebauer, Anne Birgitte, et al., eds. *Momentualising Life in the Neolithic: Narratives of Change and Continuity*. Oxford: Oxbow Books, 2020.
Geertz, Clifford. *The Interpretation of Cultures*. New York: Basic Books, 1973.
Gehl, Jan. *Life Between Buildings: Using Public Space*. Washington: Island Press, 2011.
Geldard, Richard G. *Remembering Heraclitus*. London: Lindisfarne Books, 2000.
General Authority for Statistics, Kingdom of Saudi Arabia. "Hajj Statistics: 2019–1440." Accessed Feb 2, 2023. https://www.stats.gov.sa/sites/default/files/haj_40_en.pdf.
Gertcyk, Olga. "Ancient Mummy 'with 1100 year old Adidas boots' died after she was struck on the head." *The Siberian Times*, April 12, 2017.

Gibbon, Edward. *The Decline and Fall of the Roman Empire,* edited by Antony Lentin and Brian Norman. Ware, UK: Wordsworth Editions, 1998.

Gilbreth, Frank Bunker, and Lillian Moller Gilbreth. *Fatigue Study: The Elimination of Humanity's Greatest Unnecessary Waste, a First Step in Motion Study.* New York: Macmillan, 1919.

Gladwin, Thomas. *East Is a Big Bird: Navigation and Logic on Puluwat Atoll.* Cambridge, MA: Harvard University Press, 1970.

Goddard, Tara Beth. *Drivers' Attitudes and Behaviors Toward Bicyclists: Intermodal Interactions and Implications for Road Safety—Final Report.* Portland, OR: National Institute for Transportation and Communities, 2017.

Goebel, Ted, et al. "The late Pleistocene dispersal of modern humans in the Americas." *Science* 319, no. 5869 (Mar 2008): 1497–1502.

Golding, William. *Lord of the Flies.* London: Penguin Classics, 2010.

Gordon, Aaron. "Do You Live in the US and Want a Small, Affordable Car? Too Bad." *Vice,* July 21, 2023. Accessed Apr 9, 2023. https://www.vice.com/en/article/k7qnz3/us-automakers-phasing-out-affordable-subcompact-cars.

Gössling, Stefan, and Andreas Humpe. "The global scale, distribution and growth of aviation: Implications for climate change." *Global Environmental Change* 65 (2020): 102194.

Grandin, Greg. *Fordlandia: The Rise and Fall of Henry Ford's Forgotten Jungle City.* New York: Metropolitan Books, 2009.

Greenblatt, Stephen. *The Swerve: How the World Became Modern.* New York: W.W. Norton & Company, 2011.

Greene, Kate. "Planet Boredom." *Aeon,* Feb, 26, 2014. Accessed May 17, 2023. https://aeon.co/essays/what-four-months-on-mars-taught-me-about-boredom.

Greenwood, Veronica. "Beyond." *Aeon,* August 12, 2015. Accessed Feb 11, 2023. https://aeon.co/essays/what-drives-the-urge-to-explore-the-farthest-human-reaches.

Gregory, Lady. *Visions and Beliefs in the West of Ireland.* Toronto: Colin Smythe Ltd., 1976.

Gribbin, John. *The Time Illusion.* Amazon Kindle Ed., 2016.

Gros, Frédéric. *A Philosophy of Walking.* London: Verso, 2014.

Gurdasani, Deepti, et al. "The African Genome Variation Project Shapes Medical Genetics in Africa." *Nature* 517 (2015): 327–32.

Guthrie, W.K.C. *The Greek Philosophers: from Thales to Aristotle.* London: Routledge, 2013.

______. *A History of Greek Philosophy.* Cambridge: Cambridge University Press, 1962.

Haas, Randall, et al. "Female hunters of the early Americas." *Science Advances* 6, 45 (2020).

Hadland, Tony, and Hans-Erhard Lessing. *Bicycle Design: An Illustrated History.* Cambridge: The MIT Press, 2014.

Hall, Carl. "Walk Before You Talk." *Science* 292, no. 5526 (June 2001): 2429.

Hall, Peter. *Cities of Tomorrow: An Intellectual History of Urban Planning and Design Since 1880.* Chichester: Wiley Blackwell, 2014.

Hanink, Johanna. "Even the Ancient Greeks Thought Their Best Days Were History." *Aeon,* June, 26, 2017. Accessed July 11, 2023. https://aeon.co/ideas/even-the-ancient-greeks-thought-their-best-days-were-history.

Hanlon, Sheila. "Imperial Bicyclists: Women travel writers on wheels in the late nineteenth and early twentieth century world." Accessed Oct 13, 2022. http://www.sheilahanlon.com/?p=1343.

Hann, Michael. "Why I Hate Cyclists." *The Guardian,* Oct 25, 2002.

Hanna, Judith Lynne. "To Dance Is Human." In *The Anthropology of the Body,* edited by John Blacking, 211–32. London: Academic Press, 1977.

Harari, Yuval Noah. *Sapiens: A Brief History of Humankind.* London: Vintage, 2011. Kindle.

Harcourt-Smith, William. "The First Hominins and the Origins of Bipedalism." *Evolution Education and Outreach* 3 (2010): 333–40.

Harmond, Richard. "Progress and Flight: An Interpretation of the American Cycle Craze of the 1890s." *Journal of Social History* 5, no. 2 (1971–1972): 235–57.

Hart, James A. "Du Pont General Motors Case." *Vanderbilt Law Review* 11, no. 389 (1958): 389–98.

Harvey, David. *Rebel Cities: From the Right to the City to the Urban Revolution*. London: Verso, 2012.
Hatcher, Bill. "Why You Move the Way You Do." *National Geographic*, Feb 28, 2016.
Hedges & Company. "How Many Cars Are There in the World in 2023?" Accessed Mar 5, 2023. https://hedgescompany.com/blog/2021/06/how-many-cars-are-there-in-the-world/.
Heine, Steven, and Dale S. Wright, eds. *The Kōan: Texts and Contexts in Zen Buddhism*. New York: Oxford University Press, 2000.
Heinrich, Bernd. *Why We Run: A Natural History*. New York: HarperCollins, 2001.
Heisenberg, Werner. *Encounters with Einstein and Other Essays on People, Places, and Particles*. Princeton, NJ: Princeton University Press, 1983.
The Henry Ford Website. Accessed Mar 14, 2023. https://www.thehenryford.org/explore/blog/fords-five-dollar-day/.
Heraclitus. *Fragments*. Translated by Brooks Haxton. London: Penguin Classics, 2001.
Hermanussen, Michael. "Stature of early Europeans." *Hormones: International Journal of Endocrinology and Metabolism* 2, no. 3 (2003):175–8.
Heyerdahl, Thor. *The Kon-Tiki Expedition*. London: Penguin, 1967.
Hilliard, Mark, and Áine McMahon. "Ryanair's Michael O'Leary says cyclists should be shot." *The Irish Times*, May 4, 2016.
Hobbes, Thomas. *Leviathan*. Oxford: Oxford University Press, 1966.
Hodder, Ian. "Çatalhöyük: The Leopard Changes Its Spots. A summary of recent work." *Anatolian Studies* 64 (2014): 1–22.
______. *Entangled: An Archaeology of the Relationships between Humans and Things*. Chichester: Wiley Blackwell, 2012.
______, ed. *Archaeological Theory Today*. London: Blackwell, 2001.
Holck, P. "The Oseberg ship burial, Norway: New thoughts on the skeletons from the grave mound." *European Journal of Archaeology* 9, no. 2–3 (2006): 185–210.
The Holy Teaching of Vimalakirti: A Mahayana Scripture. Translated by Robert A.F. Thuman. Philadelphia: Penn State University Press, 1976.
Horga, Laura Maria, et al. "Can marathon running improve knee damage of middle-aged adults? A prospective cohort study." *BMJ Open Sports & Exercise Medicine* 16, no. 5 (2019 Oct): e000586.
Hoskins, Tansy E. *Foot Work: What Your Shoes Tell You About Globalisation*. London: Weidenfeld & Nicolson, 2020. Kindle.
Howard, Ebenezer. *Garden Cities of Tomorrow*. Frankfurt am Main: Outlook Verlag GmbH, 2020.
Huang, Chao, et al. "Evidence of Fire Use by Homo erectus pekinensis: An XRD Study of Archaeological Bones From Zhoukoudian Locality 1, China." *Frontiers in Earth Science* 9 (2022).
Husserl, Edmund. *Shorter Works*. Notre Dame: UNDP, 1981.
I Ching or Book of Changes. Translated by Richard Wilhelm. London: Arkana, 1989.
Ingold, Tim. *Being Alive: Essays on Movement, Knowledge and Description*. Oxon, UK: Routledge, 2011.
______. *Hunters, Pastoralists and Ranchers: Reindeer Economics and Their Transformations*. Cambridge: Cambridge University Press, 1980.
International Transport Forum. "Automated and Autonomous Driving: Regulation Under Uncertainty." Paris: OECD, 2015.
Iriarte, J., et al. "Geometry by Design: Contribution of Lidar to the Understanding of Settlement Patterns of the Mound Villages in SW Amazonia." *Journal of Computer Applications in Archaeology* 3, no. 1 (2020): 151–69.
Irwin, Geoffrey. *The Prehistoric Exploration and Colonisation of the Pacific*. Cambridge: Cambridge University Press, 1992.
Jabr, Ferris. "Why Walking Helps us Think." *The New Yorker*, Sep 3, 2014.
Jacobs, Jane. *The Death and Life of Great American Cities*. New York: Vintage Books, 1961.
Jacobs, Jordan. *Taoism: A Friendly Beginners Guide on Taoism and Taoist Beliefs*. Relentless Progress Publishing, 2015. Kindle.

James, Leon, and Diane Nahl. *Road Rage and Aggressive Driving: Steering Clear of Highway Warfare.* New York: Prometheus Books, 2000.
Jarus, Owen. "Viking Ship and Cemetery Found Buried in Norway." *Livescience,* Oct 15, 2018. Accessed Feb 3, 2023. https://www.livescience.com/63829-viking-ship-cemetery.html.
Jaspers, Karl. *The Origin and Goal of History.* New Haven: Yale University Press, 1953.
Jaynes, Julian. *The Origin of Consciousness in the Breakdown of the Bicameral Mind.* Boston: Mariner, 2000.
Jenkins, Nancy. *The Boat Beneath the Pyramid: King Cheop's Royal Ship.* New York: Holt, Rinehart and Winston, 1980.
Jin, GuiYun, et al. "An important military city of the Early Western Zhou Dynasty: Archaeobotanical evidence from the Chenzhuang site, Gaoqing, Shandong Province." *China Science Bulletin* 57, nos. 2–3 (2012): 253–60.
Jones, Janet L. "Becoming a Centaur." *Aeon,* Jan 14, 2022. Accessed Mar 7, 2023. https://aeon.co/essays/horse-human-cooperation-is-a-neurobiological-miracle.
______. *Horse Brain, Human Brain: The Neuroscience of Horsemanship.* North Pomfret, VT: Trafalgar Square Books, 2020. Ebook.
Jung, C.G. *Modern Man in Search of a Soul.* New York: Harvest, 1933.
Kageyama, Peter. *For the Love of Cities: The Love Affair Between People and Their Places.* St. Petersburg: Creative Cities Productions, 2011.
Kahn, Charles H. *The Art and Thought of Heraclitus: A New Arrangement and Translation of the Fragments with Literary and Philosophical Commentary.* Cambridge: Cambridge University Press, 1979.
Kahn, Kimberly, et al. *Racial Bias in Drivers' Yielding Behavior at Crosswalks: Understanding the Effect.* Portland, OR: Transportation Research and Education Center, 2017.
Karmin M., et al. "A recent bottleneck of Y chromosome diversity coincides with a global change in culture." *Genome Research* 25, no. 4 (Apr 2015): 459–66.
Katz, Victor J. *A History of Mathematics: An Introduction.* Boston: Addison-Wesley, 2009.
Kelly, John. "Please stop riding your bike on the sidewalk in D.C.'s central business district." *The Washington Post,* Jul 7, 2014.
Kelly, Kevin. *What Technology Wants.* New York: Viking, 2010.
Kenkō and Chōmei. *Essays in Idleness.* Translated by Meredith McKinney. London: Penguin, 2013.
Kim, M.H., et al. "Reducing the frequency of wearing high-heeled shoes and increasing ankle strength can prevent ankle injury in women." *International Journal of Clinical Practice* 69, vol. 8 (Jun 2015): 909–10.
Kirk, G.S., et al. *The Presocratic Philosophers: A Critical History with a Selection of Texts.* Cambridge: Cambridge University Press, 2007.
Kitto, H.D.F. *The Greeks.* London: Penguin Books, 1951.
Knowles, J., et al. *Collisions involving pedal cyclists on Britain's roads: Establishing the causes.* Wokingham, UK: Transport Research Laboratory, 2009.
Kramer, Samuel Noah. *The Sumerians: Their History, Culture, and Character.* Chicago: University of Chicago Press, 1963.
Krause-Kyora, Ben, et al. "Neolithic and medieval virus genomes reveal complex evolution of hepatitis B." *eLife* 7 (2018).
Krishnamurti, Jiddu. *Krishnamurti's Notebook.* London: Harper & Row, 1976.
Kunstler, John Howard. *The Geography of Nowhere: The Rise and Decline of America's Man-Made Landscape.* New York: Simon & Schuster, 1993.
Kushi, Michio, and Phillip Jannetta. *Macrobiótica y Medicina Oriental.* Maldonado, Uruguay: Publicaciones GEA, 1992.
Kwitny, Jonathan. "The Great Transportation Conspiracy." *Harper's* 262, no. 1569 (1981): 14–21.
Landon, John. *Enigma of the Axial Age: History, Evolution and the Macro Effect.* Montauk, NY: South Fork Books, 2016. Kindle.
Landsburg, Steven E. *The Armchair Economist: Economics and Everyday Life.* London: Simon & Schuster, 2012.

Lao Tzu. *Tao Te Ching: The Book of the Way*. Translated by Sam Torode. Nashville: Ancient Renewal, 2018.
______. *Tao Te Ching: The Classic Book of Integrity and the Way*. Translated by Victor H. Mair. New York: Bantam Books, 1990.
Larminie, James, and John Lowry. *Electric Vehicle Technology Explained*. Chichester: John Wiley & Sons Ltd., 2012.
Lazare, Sarah. "Meet the Yemeni Woman Using Creative Direct Action to Resist the Country's Brutal War." Accessed Mar 15, 2016. https://www.alternet.org/world/meet-yemeni-woman-using-creative-direct-action-resist-countrys-brutal-war?fbclid=IwAR2swHuL4bpg1ABkpFdU2_X6alXwqsCpgUxB5p7usuRDiF7oIFqebHTlaUI.
Lazaridis, Iosif, et al. "A genetic probe into the ancient and medieval history of Southern Europe and West Asia." *Science*, 377 (2022): 940–51.
Le Corbusier. *The City of Tomorrow and Its Planning*. New York: Dover Publications, Inc., 1987.
Lefebvre, Henri. *The Production of Space*. Oxford: Blackwell, 1991.
______. *Writings on Cities*. Oxford: Blackwell, 1996.
Lessing, Hans-Erhard: "The evidence against Leonardo's bicycle." *Cycle History* 8 (1998): 49–56.
Lévi-Strauss, Claude. *The Savage Mind*. London: Weidenfeld and Nicolson, 1966.
Lewis-Williams, J.D. "Quanto?: The Issue of 'Many Meanings' in Southern African San Rock Art Research." *The South African Archaeological Bulletin* 53, no. 168 (1998): 86–97.
Leyland, Louise-Ann, et al. "The effect of cycling on cognitive function and well-being in older adults." *PloS One* 14, no. 2 (2019).
Liebenberg, Louis. "The Relevance of Persistence Hunting to Human Evolution." *Journal of Human Evolution* 55, no. 6 (2008): 1156–9.
Lieberman, Daniel. *The Story of the Human Body: Evolution, Health, and Disease*. New York: Pantheon Books, 2013.
______ et al. "The evolution of endurance running and the tyranny of ethnography: A reply to Pickering and Bunn (2007)." *Journal of Human Evolution* 53 (2007): 439–42.
Lindner, Stephan. "Chariots in the Eurasian Steppe: a Bayesian approach to the emergence of horse-drawn transport in the early second millennium BC." *Antiquity* 94, no. 374 (2020): 361–80.
Littlejohn, Ronnie L. *Confucianism: An Introduction*. London: I.B. Tauris, 2011.
Lizier, Daniele S., et al. "Effects of Reflective Labyrinth Walking Assessed Using a Questionnaire." *Medicines* 5, no. 4 (2018): 111.
Loewe, Michael, and Edward L. Shaughnessy, eds. *The Cambridge History of Ancient China: From the Origins of Civilization to 221 BC*. Cambridge: Cambridge University Press, 1999.
Lohner, Svenja. "Sensing with your Feet!" *Scientific American*, July 20, 2017.
The Long Discourses of the Buddha: A Translation of the Dīgha Nikāya. Translated by Maurice Walshe. Boston: Wisdom Publications, 1995.
Lucibello, K.M., et al. "Examining a training effect on the state anxiety response to an acute bout of exercise in low and high anxious individuals." *Journal of Affective Disorders* 247 (Dec 2019): 29–35.
Luckert, Erika. "Drawings We Have Lived: Mapping Desire Lines in Edmonton." *Constellations* 4, no. 1 (2013): 318–27.
Lucretius. *On the Nature of Things*. Translated by William Ellery Leonard. Global Grey, 2018. Kindle.
MacFadden, Bruce J., and Robert P. Guralnick. "Horses in the Cloud: Big data exploration and mining of fossil and extant Equus (Mammalia: Equidae)." *Paleobiology* 43, no. 1 (2016): 1–14.
Mack, John. *The Sea: A Cultural History*. London: Reaktion Books Ltd., 2011.
Mack, Katie. *The End of Everything (Astrophysically Speaking)*. New York: Simon & Schuster, 2020. Kindle.
Marlowe, Frank W. "Hunter-Gatherers and Human Evolution." *Evolutionary Anthropology* 14, no. 2 (Apr 2005): 54–67.

Marsal, Adam, and Brian Fleming. "Riding for your Rights!" *We Love Cycling*, Jun 10, 2015. Accessed Jun 15, 2023. https://www.welovecycling.com/wide/2015/06/10/riding-for-your-rights/.

Marshall, Colin. "Story of Cities #29: Los Angeles and the 'great American streetcar scandal.'" *The Guardian*, Apr 25, 2016. Accessed Jun 15, 2023, https://www.theguardian.com/cities/2016/apr/25/story-cities-los-angeles-great-american-streetcar-scandal.

Marshall, Michael. "Supervolcano eruptions may not be so deadly after all." *New Scientist*, Apr 29, 2013. Accessed Jul 12, 2023. https://www.newscientist.com/article/dn23458-supervolcano-eruptions-may-not-be-so-deadly-after-all/.

Marx, Karl, and Friedrich Engels. *The Communist Manifesto*. London: Penguin Books, 1967.

Massumi, Brian. *Parables for the Virtual: Movement, Affect, Sensation*. Durham: Duke University Press, 2002.

Matthews, W.H. *Mazes and Labyrinths: A General Account of Their History and Developments*. Glastonbury: The Lost Library, 1922.

Matthiessen, Peter. *The Snow Leopard*. London: Vintage, 2003.

McDougall, Ian, et al. "Stratigraphic placement and age of modern humans from Kibish, Ethiopia." *Nature* 433 (2005): 733–6.

McGilchrist, Iain. *The Master and His Emissary: The Divided Brain and the Making of the Western World*. New Haven: Yale University Press, 2018.

McLuhan, Marshall. *Understanding Media: The Extensions of Man*. Cambridge: MIT Press, 1994.

McNeill, William H. *Keeping Together in Time: Dance and Drill in Human History*. Cambridge, MA: Harvard University press, 1995. Kindle.

McShane, Clay, and Joel A. Tarr. *The Horse in the City: Living Machines in the Nineteenth Century*. Baltimore: Johns Hopkins University Press, 2007.

Melé, Marta, et al. "The Genographic Consortium, Recombination Gives a New Insight in the Effective Population Size and the History of the Old World Human Populations." *Molecular Biology and Evolution* 29, no. 1 (Jan 2012): 25–30.

Melville, Herman. *Moby Dick or The Whale*. New York: Penguin, 2003.

Mencius. *Mencius*. Translated by D.C. Lau. London: Penguin Classics, 2003.

Merleau-Ponty, Maurice. *Phenomenology of Perception*. Translated by Colin Smith. London: Routledge, 2002.

Merriman, Peter. *Mobility, Space and Culture*. London: Routledge, 2012.

The Middle Length Discourses of the Buddha: A Translation of the Majjhima Nikāya. Translated by Bhikkhu Ñanamole and Bhikkhu Bodhi. Boston: Wisdom Publications, 1995.

Miller, Ross H., and Rebecca L. Krupenevich. "Medial knee cartilage is unlikely to withstand a lifetime of running without positive adaptation: a theoretical biomechanical model of failure phenomena." *PeerJ* 8 (2020): 1–27.

Miller, Ross H., et al. "Why don't most runners get knee osteoarthritis? A case for per-unit-distance loads." *Medicine & Science in Sports & Exercise* 46, no. 3 (Mar 2014): 572–9.

Milloy, Courtland. "Bicyclist bullies try to rule the road in D. C." *The Washington Post*, July 8, 2014.

Milner, George R. "Early agriculture's toll on human health." *PNAS* 116, no. 28 (Jul 2019): 13721–3.

Mithen, Steven. *The Prehistory of the Mind: A Search for the Origins of Art, Religion and Science*. London: Thames & Hudson, 1996.

______, ed. *Creativity in Human Evolution and Prehistory*. London: Routledge, 2005.

Mittermaier, Amira. *Dreams That Matter: Egyptian Landscapes of the Imagination*. Berkeley, CA: University of California Press, 2011.

Mo Zi. *The Book of Master Mo*. Translated by Ian Johnston. London: Penguin Classics, 2013. Kindle.

Montaigne, Michel de. *The Complete Works: Essays, Travel Journal, Letters*. Translated by Donald M. Frame. New York: Alfred A. Knopf, 2003.

Morris, Desmond. *The Naked Ape: A Zoologist's Study of the Human Animal*. London: Jonathan Cape, 1967.

Morris, James. *Cities*. London: Faber & Faber, 1963.
Mumford, Lewis. *The City in History: Its Origins, Its Transformations, and Its Prospects*. New York: Harcourt, Brace & World Inc., 1961.
______. *The Culture of Cities*. San Diego: Harcourt Brace Jovanovich Publishers, 1970.
______. *The Story of Utopias*. Azafran Books, 2017.
______. *Technics & Civilization*. Chicago: University of Chicago Press, 2010.
Murphy, Dervla. *Wheels Within Wheels: The Making of a Traveller*. London: Eland Publishing, 2011. Kindle.
Murti, T.R.V. *The Central Philosophy of Buddhism*. London: Routledge, 2008.
Myers, William A. *Trolleys to the Surf: The Story of the Los Angeles Pacific Railway*. Glendale: Interurban Press, 1976.
Nader, Ralph. *Unsafe at Any Speed: The Designed-In Dangers of the American Automobile*. London: Knightsbridge, 1991.
Nail, Thomas. *Being and Motion*. New York: Oxford University Press, 2019.
______. *The Figure of the Migrant*. Stanford, CA: Stanford University Press, 2015. Kindle.
______. *Lucretius I: An Ontology of Motion*. Edinburgh: Edinburgh University Press, 2018.
Napier, John. *The Antiquity of Human Walking*. New York: W.H. Freeman and Company, 1967.
______. "The Antiquity of Human Walking." *Scientific American* 216, no. 4 (Apr 1967): 467–56.
Needham, Andy, et al. "Art by firelight? Using experimental and digital techniques to explore Magdalenian engraved plaquette use at Montastruc (France)." *PloS One* 17, no 4 (Apr 2022): 1–28.
Nemet-Nejat, Karen Rhea. *Daily Life in Ancient Mesopotamia*. Westport, CT: Greenwood Press, 1998.
Neumann, Erich. *The Origin and History of Consciousness*. Oxon, UK: Routledge, 2002.
Nichols, Wallace J. *Blue Mind: The Surprising Science That Shows How Being Near, In, On, or Under Water Can Make You Happier, Healthier, More Connected, and Better at What You Do*. London: Abacus, 2018.
Nicholson, Geoff. *The Lost Art of Walking: The History, Science, Philosophy, and Literature of Pedestrianism*. London: Riverhead Books, 2008.
Nipper, Sarah. "Wheels of Change: The Bicycle and Women's Rights." *MS Magazine*, May 7, 2014.
Nock, O.S. *World Atlas of Railways*. Bristol: Victoria House Publishing, 1983.
Norager, Joshua. *Nonduality: A Brief Introduction*. Self-Published, 2015. Kindle.
Norcliffe, Glen. *The Ride to Modernity: The Bicycle in Canada, 1869–1900*. Toronto: University of Toronto Press, 2001.
Noss, Andrew J., and Barry S. Hewlett. "The Contexts of Female Hunting in Central Africa." *American Anthropologist* 103, no. 4 (2001): 1024–40.
Núñez, Rafael E., and Eve Sweetser. "With the Future Behind Them: Convergent Evidence from Aymara Language and Gesture in the Crosslinguistic Comparison of Spatial Construals of Time." *Cognitive Science* 30, no. 3 (2006): 401–50.
Olalde, Íñago, et al. "The genomic history of the Iberian Peninsula over the past 8000 years." *Science* 363, no. 6432 (Mar. 2019): 1230–4.
Online Etymological Dictionary. Accessed May 9, 2023. https://www.etymonline.com/.
Oppenheimer, Stephen. *Out of Eden: The Peopling of the World*. London: Robinson, 2003. Kindle.
Ord-Hume, Arthur W.J.G. *Perpetual Motion: The History of an Obsession*. New York: St. Martin's Press, 1977.
Ovid. *Metamorphoses: A New Verse Translation*. Translated by David Raeburn. London: Penguin Classics, 2004. Ebook.
Paine, Lincoln. *The Sea and Civilisation: A Maritime History of the World*. New York: Alfred A. Knopf, 2013.
Pakendorf, Brigitte, and Mark Stoneking. "Mitochondrial DNA and Human Evolution." *Nature* 325 (1987): 31–6.
Palaephatues: *On Unbelievable Tales*. Translated by Jacob Stern. Wauconda, IL: Bolchazy-Carducci Publishers, 1996.

Parissien, Steven. *The Life of the Automobile: A New History of the Motor Car.* London: Atlantic Books, 2013.
Parris, Matthew. "What's Smug and Deserves to be Decapitated?" *The Times,* Dec 27, 2007.
Paulus, Nathan. "Car Ownership Statistics in the US." MoneyGeek. Accessed Apr 5, 2023. https://www.moneygeek.com/insurance/auto/car-ownership-statistics/#vehicle-ownership-by-city.
Peden, Margie, et al., eds. *World Health Organization World Report on Road Traffic Injury Prevention.* Geneva, 2004.
Peltzman, Sam. "The Effects of Automobile Safety Regulation." *Journal of Political Economy* 83, no. 4 (Aug 1975): 677–726.
Pennell, Elizabeth Robins. *Over the Alps on a Bicycle.* Charlotte: Eltanin Publishing, 2014. Kindle.
"The Perilous Politics of Parking." *The Economist,* Apr 6, 2017.
Piketty, Thomas. *Capital in the Twenty-First Century.* Cambridge: Belknap Press, 2014.
Pinto, Nataly. *Mujeres en Bici: Una Expersión de Libertad que Transciende Fronteras.* Quito: FES-ILDIS, 2017.
Plato. *The Last Days of Socrates: Euthyphro, Apology, Crito, Phaedo.* Translated by H. Tredennick and H. Tarrant. London: Penguin Books, 1993.
______. *Symposium.* Translated by M.C. Howatson. Cambridge: Cambridge University Press, 2008.
Polkinghorne, John. *Quantum Theory: A Very Short Introduction.* Oxford: Oxford University Press, 2002.
Population Reference Bureau. "2016 World Population Fact Sheet." Accessed Jun 15, 2023. www.prb.org.
Posamentier, Alfred S., and Ingmar Lehmann. *Pi: A Biography of the World's Most Mysterious Number.* New York: Prometheus Books, 2004.
Powell, Eric. "Oldest Egyptian Funerary Boat." *Archaeology: A Publication of the Archaeology Institute of America.* Jan/Feb 2013.
Price, Huw. *Time's Arrow and Archimedes' Point: New Directions in the Physics of Time.* Oxford: Oxford University Press, 1996.
Pross, Addy. "Life's restlessness." *Aeon,* Apr 29, 2014. Accessed May 19, 2023. https://aeon.co/essays/paradoxes-of-stability-how-life-began-and-why-it-can-t-rest.
Quinn, Bob. *The Atlantean Irish: Ireland's Oriental and Maritime Heritage.* Dublin: Lilliput Press, 2005.
Radkau, Joachim. *Nature and Power: A Global History of the Environment.* Cambridge: Cambridge University Press, 2002.
Rampino, Michael, and Stephen Self. "Bottleneck in Human Evolution and the Toba Eruption." *Science* 262, no. 5142 (1993): 1955.
Rantala, M.J. "Evolution of Nakedness in *Homo sapiens.*" *Journal of Zoology* 273, no. 4 (2007): 279.
Ratchnevsky, Paul. *Genghis Khan: His Life and Legacy.* Translated by Thomas Nivison Haining. Oxford: Blackwell, 1991.
Rau, A.R.P. "The Rosetta-Philae Comet Mission as Physics Appreciation." *Resonance* 20 (4) (Apr 2014): 346–51.
Reader, John. *Cities.* London: Vintage, 2005.
Reed, Drew. "Fordlandia—The Failure of Henry Ford's Utopian City in the Amazon." *The Guardian,* Aug 19, 2016.
Reid, Carlton. *Roads Were Not Built for Cars: How Cyclists Were the First to Push for Good Roads & Became the Pioneers of Motoring.* Washington, D.C.: Island Press, 2015.
Relph, Edward. *Place and Placelessness.* London: Pion Ltd., 1976.
Renfrew, Colin. "Neuroscience, Evolution and the Sapient Paradox: The Factuality of Value and of the Sacred." *Philosophical Transactions of the Royal Society London, Biological Sciences,* Jun 12, 2008.
______. "The Sapient Paradox: Social Interaction as a Foundation of Mind." Nov 14, 2016. Accessed Feb 23, 2023. https://www.youtube.com/watch?v=Xgl7b02Ub6Y&list=PLUmOb2SQEIwlfEb6KEP7dy6T-37YqPN45&index=5.

Riess, Adam G., et al. "Observational Evidence from Supernovae for an Accelerating Universe and a Cosmological Constant." *The Astronomical Journal.* 116, no. 3 (1998): 1009–38.
Rindos, David. *The Origins of Agriculture: An Evolutionary Perspective.* San Diego: Academic Press Inc., 1984.
Roberts, Alice. *The Incredible Human Journey: The Story of How We Colonised the Planet.* London: Bloomsbury, 2009. Kindle.
Robinson, Tim. *My Time in Space.* Dublin: Lilliput Press, 2001.
Roe, Jeremy. *Antoni Gaudí.* New York: Parkstone International, 2019.
Rolian, Campbell, et al. "Walking, Running and the Evolution of Short Toes in Humans." *Journal of Experimental Biology* 212, no. 5 (2009): 713–21.
Ronto, Paul. "The State of Ultra Running 2020." Run Repeat. Accessed Feb 6, 2023. https://runrepeat.com/state-of-ultra-running. September 21, 2021.
Rouse Ball, W.W. *Mathematical Recreations and Essays.* London: Macmillan, 1905.
Rousseau, Jean-Jacques. *The Confessions.* Feedbooks [1768]. Kindle.
______. *Reveries of the Solitary Walker.* Translated by Peter France. London: Penguin Books, 1979.
Rovelli, Carlo. *Reality Is Not What It Seems: The Journey to Quantum Gravity.* London: Penguin, 2016. Kindle.
Rudebusch, George. "Dramatic Prefiguration in Plato's *Republic.*" *Philosophy and Literature* 26, no. 1 (Apr 2002): 75–83.
Sachs, Curt. *World History of the Dance.* Translated by Bessie Schönberg. New York: W.W. Norton & Company, 1937.
Sagan, Carl. *Pale Blue Dot: A Vision of the Human Future in Space.* New York: Ballantine Books, 1994.
Sagona, Claudia. *The Archaeology of Malta.* New York: Cambridge University Press, 2015.
Sahlins, Marshall. *Stone Age Economics.* Chicago: Aldine-Atherton Inc., 1972.
Sangharakshita. *The Bodhisattva Ideal.* Cambridge: Windhorse, 1999.
______. *Complete Works.* vol 3. Cambridge: Windhorse Publications, 2017.
______. *A Survey of Buddhism: Its Doctrines and Methods Through the Ages.* Birmingham: Windhorse, 2001.
Sargent, Lyman Tower. *Utopianism: A Very Short Introduction.* Oxford: Oxford University Press, 2010.
Sattin, Anthony. *Nomads: The Wanderers Who Shaped Our World.* London: John Murray Publishers Ltd., 2022.
Saward, Jeff. *Labyrinths & Mazes: A Complete Guide to Magical Paths of the World.* Asheville, NC: Lark Books, 2003.
Sayer, Amber. "How many people have run a marathon: World Statistics." *Marathonhandbook,* Nov 18, 2022. Accessed July 10, 2023. https://marathonhandbook.com/how-many-people-have-run-a-marathon/#:~:text=According%20to%20the%20US%20Census,population%20has%20run%20a%20marathon.
Scham, Sandra. "The World's First Temple." *Archaeology* 61, no. 6 (Nov/Dec 2008): 22–7.
Schivelbusch, Wolfgang. *The Railway Journey: The Industrialization of Time and Space in the Nineteenth Century.* Oakland, CA: University of California Press, 2014. Kindle.
Schmitt, Daniel, et al. "Experimental Evidence Concerning Spear Use in Neanderthals and Early Modern Humans." *Journal of Archaeological Science* 30 (2003): 103–14.
Searle, John R. *Intentionality: An Essay in the Philosophy of Mind.* Cambridge: Cambridge University Press, 1983.
Sears, Edward S. *Running Through the Ages.* Jefferson, NC: McFarland, 2015.
Seddon, Christopher. *Humans: From the Beginning: From the First Apes to the First Cities.* London: Glanville Publications, 2014.
Semmelhack, Elizabeth. *Shoes: The Meaning of Style.* London: Reaktion Books, 2017.
Seneca. *Letters from a Stoic: Epistulae Morales ad Lucilium.* Translated by Robin Campbell. London: Penguin Books, 1969.
Sennett, Richard. *The Fall of Public Man.* New York: W.W. Norton & Company, 1976. Kindle.
Setright, L.J.K. *Drive On! A Social History of the Motor Car.* London: Granta Books, 2004.

Seyfzadeh, M., and R. Schoch. "World's First Known Written Word at Göbekli Tepe on T-Shaped Pillar 18 Means God." *Archaeological Discovery* 7, no. 2 (Apr 2019): 31–53.
Sharp, N.C.C. "Timed running speed of a cheetah (Acinonyx jubatus)." *Journal of Zoology* 241, no. 3 (Mar 1997): 493–4.
Shaw, Ben, et al. "Emergence of a Neolithic in highland New Guinea by 5000 to 4000 years ago." *Science Advances* 6, no. 13 (2020).
Sheehan, George. *Running & Being: The Total Experience.* Emmaus, PA: Rodale, 1978.
Sheffi, Yosef, and Carlos F. Daganzo. "Another 'paradox' of traffic flow." *Transportation Research* 12, no. 1 (1978): 43–6.
Shostak, Marjorie. *Nisa: The Life and Words of a !Kung Woman.* Cambridge, MA: Harvard University Press, 1981.
Shoup, Donald C. "The trouble with minimum parking requirements." *Transportation Research Part A: Policy and Practice* 33, no. 7–8 (1999): 549–74.
Simmons, Alan H. *Stone Age Sailors: Prehistoric Seafaring in the Mediterranean.* Walnut Creek, CA: Left Coast Press, 2014.
Skilton, Andrew. *A Concise History of Buddhism.* Cambridge: Windhorse Publications Ltd., 2013. Epub.
Slavin, Terry. "If there aren't as many women cycling as men…you need better infrastructure." *The Guardian,* Jul 9, 2015.
Smith, Sharon. "Engels and the Origin of Women's Oppression." *International Socialist Review* 2, no. 3 (1997).
Smith C., ed. *Encyclopedia of Global Archaeology.* New York: Springer, 2014.
Smith, Robin, and Kevin Heatherington, eds. *Urban Rhythms: Mobility, Space and Interaction in the Contemporary City.* Oxford: Wiley Blackwell, 2013.
Solnit, Rebecca. *A Field Guide to Getting Lost.* Edinburgh: Canongate, 2017.
______. *Motion Studies: Time, Space and Eadweard Muybridge.* London: Bloomsbury, 2003.
______. *Wanderlust: A History of Walking.* New York: Penguin, 2001. Kindle.
Sourisseau, Jean-Michel, ed. *Family Farming and the Worlds to Come.* Paris: Springer, 2015.
Spengler, Oswald. *The Decline of the West.* Petosky, MI: Random Shack, 2016.
______. *Man and Technics: A Contribution to a Philosophy of Life.* Budapest: Arktos, 2015.
Sperling, Daniel, and Deborah Gordon. "Two Billion Cars: Transforming a Culture." *TR News* 259 (Nov/Dec, 2008).
Spero, Rosanna. *RAC Report on Motoring 2012.* Walsal, UK: RAC House, 2012.
Splichal, Emily. "The effect of sensory stimulation on movement accuracy." *Lermagazine,* Mar 2018. Accessed Jan 20, 2023. https://lermagazine.com/article/the-effect-of-sensory-stimulation-on-movement-accuracy#:~:text=The%20foot%20is%201%20of,adapting%20type%202%20(FA2).
Sri Aurobindo. *Essays in Philosophy and Yoga: The Complete Works of Sri Aurobindo.* Pondicherry: Sri Aurobindo Ashram Publication Department, 1998. Kindle.
Stafford, Tom. "The psychology of why cyclists enrage car drivers." *BBC,* Feb 12, 2013. Accessed 16 Mar. 2023. https://www.bbc.com/future/article/20130212-why-you-really-hate-cyclists.
Stevens, John. *The Marathon Monks of Mount Hiei.* Boston: Shambala, 1988.
Story, Benjamin, and Jenna Silber Storey. *Why We Are Restless: On the Modern Quest for Contentment.* Princeton, NJ: Princeton University Press, 2021.
Strehlow, Theodor. *Songs of Central Australia.* Sydney: Angus and Robertson, 1978.
Stringer, C., and P. Andrews. "Genetic and Fossil Evidence for the Origin of Modern Humans." *Science* 239, no. 4845 (Mar. 1988): 1263–8.
Stringer, Chris. *Lone Survivors: How We Came to Be the Only Humans on Earth.* New York: Times Books, 2012.
Stromberg, Joseph. "'Bicycle Face': a 19th-century health problem made up to scare women away from biking." *Vox,* Mar 24, 2015. Accessed Jun 15, 2023. https://www.vox.com/2014/7/8/5880931/the-19th-century-health-scare-that-told-women-to-worry-about-bicycle.
Strootman, Rolf. "Alexander's Thessalian cavalry." *Talanta* 42/43 (2010–2011) 51–67.
Sun Tzu, *The Art of War.* Translated by Lionel Giles. Pax Librorum, 2009.

Svendsen, Lars. *A Philosophy of Boredom*. London: Reaktion Books, 2005.
Taalbi, Josef, and Hana Nielsen. "The role of energy infrastructure in shaping early adoption of electric and gasoline cars." *Nat Energy* 6 (2021): 970–6.
Tarnas, Richard. *The Passion of the Western Mind: Understanding the Ideas That Have Shaped Our World View*. New York: Ballantine Books, 1991.
Taylor, Bron, ed. *The Encyclopedia of Religion and Nature*. New York: Continuum, 2005.
Tegmark, Max. *Our Mathematical Universe: My Quest for the Ultimate Nature of Reality*. London: Penguin, 2015.
Tester, Keith, ed. *The Flâneur*. New York: Routledge, 1994.
Thomson, Russell, et al. "Recent Common Ancestry of Human Y Chromosomes: Evidence from DNA Sequence Data." *Proceedings of the National Academy of the Sciences of the United States of America* 97, no. 13 (June 2000): 7360–5.
Thorpe, I.J. *The Origins of Agriculture in Europe*. London: Routledge, 1996.
Timney, Brian, and Todd Macuda. "Vision and Hearing in Horses." *Journal of the American Veterinary Medical Association* 218, no. 10 (2001): 1567–74.
Tishkoff, Sarah A., et al. "The Genetic Structure and History of Africans and African Americans." *Science* 325, no. 5930 (2009): 1035–44.
Tolentino, Cierra. "Centaurs: Half-Horse Men of Greek Mythology." *History Cooperative*, Oct 17, 2022. Accessed March 6, 2023. https://historycooperative.org/centaurs/.
Toynbee, Arnold J. *A Study of History*, abridged ed. Oxford: Oxford University Press, 1987. Kindle.
Trautmann, Martin, et al. "First bioanthropological evidence for Yamnaya horsemanship." *Science Advances* 9, no. 9 (March 2023).
Trinkaus, Erik, and Hong Shang. "Anatomical evidence for the antiquity of human footwear: Tianyuan and Sunghir." *Journal of Archaeological Science* 35, no. 7 (2008): 1928–33.
Turner, Steven. "Meet Bushra Al-Fusail & the Yemeni women cycling in the face of oppression & civil war." *Huck Magazine*, July 26, 2015. Accessed June 15, 2023. https://www.huckmag.com/article/bushra-al-fusail.
United Nations. "Overcoming Barriers: Human Mobility and Development." United Nations Human Development Report 2009. Accessed Feb 8, 2023. http://oppenheimer.mcgill.ca/IMG/pdf/HDR_2009_EN_Complete.pdf.
United States Department of Transportation. "Federal Highway Administration." May 31, 2022. https://www.fhwa.dot.gov/ohim/onh00/bar8.htm. Accessed Mar 17, 2023.
United States District Court. "N.D. Illinois, United States v. National City Lines. 134 F. Supp. 350 (N.D. Ill. 1955)." Accessed Apr 4, 2023. https://casetext.com/case/united-states-v-national-city-lines-2.
Uomini, Natalie T. "Handedness in Neanderthals." In *Neanderthal Lifeways, Subsistence and Technology: One Hundred Fifty Years of Neanderthal Study*, edited by Nicholas J. Conard and Jürgen Richter. New York: Springer, 2010.
Urry, John. *Mobilities*. Cambridge: Polity Press, 2007.
Vanderbilt, Tom. *Traffic: Why We Drive the Way We Do (and What It Says About Us)*. New York: Alfred A. Knopf, 2008. Kindle.
Van der Post, Laurens. *The Lost World of the Kalahari*. Hammondsworth: Penguin Books, 1958.
Van Haaren, Rob. "Assessment of Electric Cars' Range Requirements and Usage Patterns based on Driving Behavior recorded in the National Household Travel Survey of 2009." *Solar Journey USA*. July 2012.
Vasari, Giorgio. *The Lives of the Artists*. Translated by Julia Conaway Bondanella and Peter Bondanella. Oxford: Oxford University Press, 1991.
Veblen, Thorstein. *The Theory of the Leisure Class*. Oxford: Oxford University Press, 2007.
Vince, Gaia. *Nomad Century: How to Survive the Climate Upheaval*. London: Allen Lane, 2022.
Virilio, Paolo. *Speed and Politics: An Essay on Dromology*. Translated by M. Polizzotti. South Pasadena, CA: Semiotext(e), 2006.
Voegelin, Eric. *Order and History*. Columbia, MO: University of Missouri Press, 1901.

Volken, Marquita. "Arming Shoes of the Fifteenth Century." *Acta Periodica Duellatorum* 5, no. 2 (2017): 25–45.
Volti, Rudi. *Cars and Culture: The Life Story of a Technology.* Baltimore: Johns Hopkins University Press, 2004.
______. *Society and Technological Change.* New York: Worth Publishers, 2014.
Wade, Nicholas. *Before the Dawn: Recovering the Lost History of Our Ancestors.* London: Penguin, 2006.
Waldron, Tony. *Shadows in the Soil: Human Bones & Archaeology.* London: Tempus, 2001.
Walker, Alan, and Chris Stringer, eds. *The First Four Million Years of Human Evolution.* London: Royal Society, 2010.
Walker, Ian. "Drivers overtaking bicyclists: Objective data on the effects of riding position, helmet use, vehicle type and apparent gender." *Accident Analysis and Prevention* 39, no. 2 (2007): 417–25.
Walker, Peter. "Sabotage and hatred: what have people got against cyclists?" *The Guardian,* Jul 1, 2015. Accessed Jun 15, 2023. https://www.theguardian.com/lifeandstyle/2015/jul/01/sabotage-and-hatred-what-have-people-got-against-cyclists.
Ward, Caroll, et al. "The new hominid species Australopithecus anamensis." *Evolutionary Anthropology: Issues, News, and Reviews* 7, no. 6 (1999): 197–205.
Ward, Maria E. *Bicycling for Ladies.* New York: Bretano's, 1896. Kindle.
Waterfield, Robin. *The First Philosophers: The Presocratics and Sophists.* Oxford: Oxford University Press, 2000.
Weatherford, Jack. *Genghis Khan and the Making of the Modern World.* New York: Broadway Books, 2005. Kindle.
White, Lynn, Jr. *Medieval Technology and Social Change.* London: Oxford University Press, 1962.
Wilkins, Van. "The Conspiracy Revisited." *The New Electric Railway Journal* (1995): 18–22.
Wilkinson, Richard, and Kate Pickett. *The Spirit Level: Why Equality Is Better for Everyone.* New York: Bloomsbury Press, 2010.
Willard, Frances. *Wheel Within a Wheel: How I Learned to Ride the Bicycle with Some Reflections by the Way.* Bedford, MA: Applewood Books, 1895. Kindle.
Williams, Wendy. *The Horse: The Epic History of Our Noble Companion.* New York: Macmillan, 2015.
Wilson, Edward O. *The Future of Life.* New York: Vintage, 2003.
Winkless, Laurie. *Science and the City: The Mechanics Behind the Metropolis.* London: Bloomsbury, 2016.
Wölfflin, Heinrich. *Renaissance and Baroque.* London: Fontana Library, 1964.
Wollen, Peter, and Joe Kerr. *Autopia: Cars and Culture.* London: Reaktion Books, 2002.
Wolmar, Christian. *Blood, Iron & Gold: How the Railways Transformed the World.* London: Atlantic, 2009.
Woodburn, James. "Egalitarian Societies." *Man* 17, no. 3 (1982): 431–51.
World Health Organization. "Physical Activity Fact Sheet." Accessed Jan 26, 2023. https://www.who.int/news-room/fact-sheets/detail/physical-activity.
______. "Road Safety Traffic Deaths." Accessed Apr 9, 2023. https://www.who.int/gho/road_safety/mortality/traffic_deaths_number/en/.
______. "World Migration Report 2010." Accessed Feb 8, 2023. https://www. iom.int/world-migration-report-2010.
Wosien, Maria-Gabriele. *Sacred Dance: Encounter with the Gods.* New York: Thames and Hudson, 1986.
Wrangham, Richard. *Catching Fire: How Cooking Made Us Human.* London: Profile Books Ltd., 2009.
Xinzhong Yao, ed. *The Encyclopedia of Confucianism.* New York: Routledge, 2003.
The Yamnaya Impact on Prehistoric Europe Project. Accessed Mar 9, 2023. https://www.helsinki.fi/en/researchgroups/the-yamnaya-impact-on-prehistoric-europe.
Yanocha, Dana. "Optimising New Mobility Services." *International Transport Forum Discussion Papers.* Paris: OECD, 2018.

Yeats, W.B. *The Collected Poems of W.B. Yeats*. Edited by Richard J. Finneran. New York: Palgrave Macmillan, 1989.

Yu, Ning, et al. "Larger Genetic Differences Within Africans Than Between Africans and Eurasians." *Genetics* 161, no. 1 (May 2002): 269–74.

Zheutlin, Peter. *Around the World on Two Wheels: Annie Londonderry's Extraordinary Ride*. New York: Citadel Press, 2007.

______. "Backstory: Chasing Annie Londonderry." *The Christian Science Monitor,* Aug 28, 2006. Accessed Jun 15, 2023. https://www.csmonitor.com/2006/0828/p20s01-algn.html.

______. "Women on Wheels: The Bicycle and the Women's Movement of the 1890s." Annielondonderry.com. Accessed Jun 18, 2018. annielondonderry.com/womenWheels/html.

Zucker, Paul. "Space and Movement in High Baroque City Planning." *Journal of the Society of Architectural Historians* 14, no. 1 (1955): 8–15.

Index

Numbers in ***bold italics*** indicate pages with illustrations